Miraflores

Mira

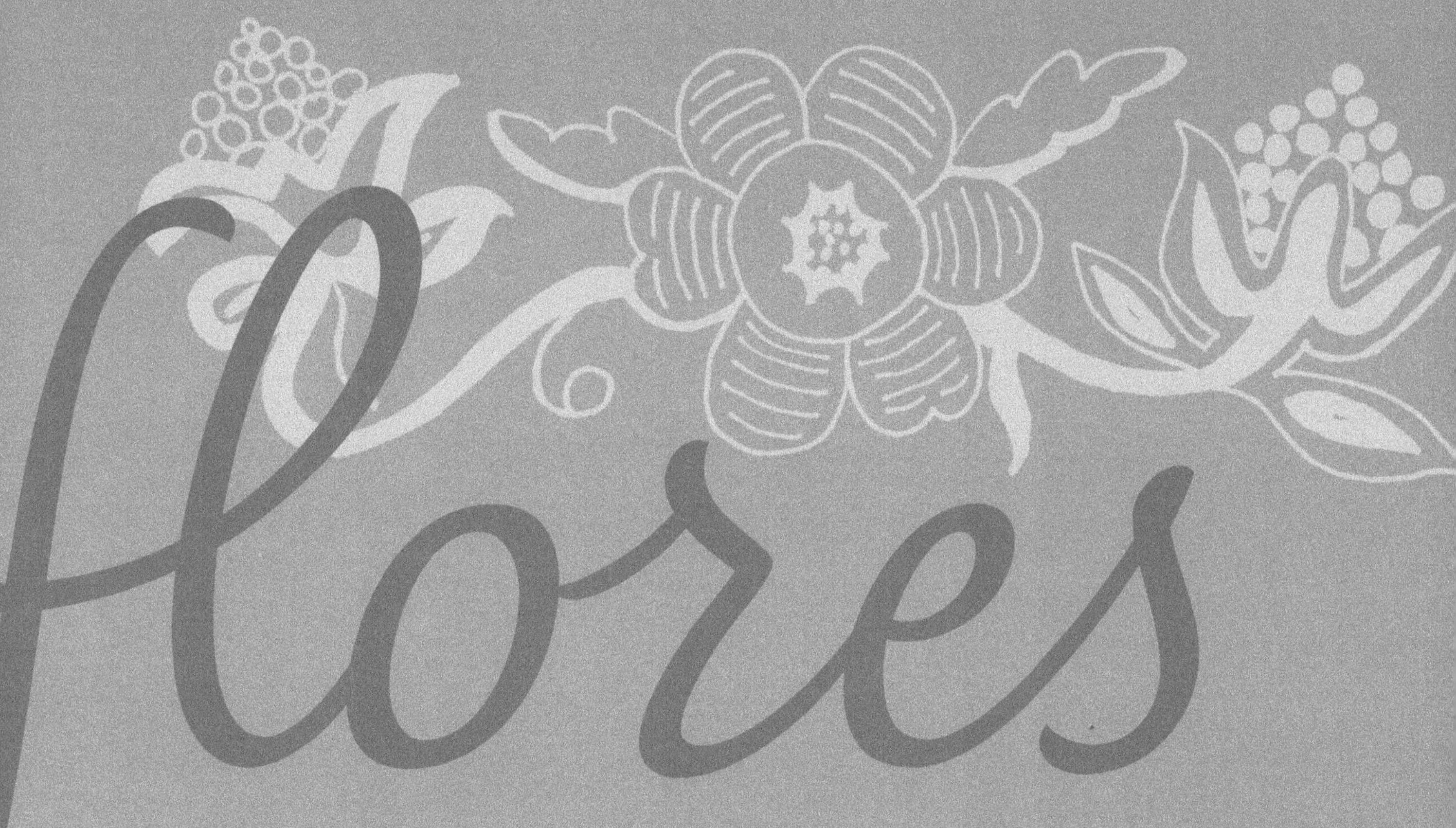

flores

San Antonio's Mexican Garden of Memory

ANNE ELISE URRUTIA

FOREWORD BY Tomás Ybarra-Frausto

MAVERICK BOOKS | TRINITY UNIVERSITY PRESS
SAN ANTONIO

Maverick Books, an imprint of
Trinity University Press
San Antonio, Texas 78212

Book design by Anne Richmond Boston
Author photo by Josh Huskin
Site illustrations by Rebecca Schenker
Talavera motif illustrations by Anne Elise Urrutia

ISBN 978-1-59534-936-1 paper
ISBN 978-1-59534-937-8 ebook

Trinity University Press strives to produce its books using methods and materials in an environmentally sensitive manner. We favor working with manufacturers that practice sustainable management of all natural resources, produce paper using recycled stock, and manage forests with the best possible practices for people, biodiversity, and sustainability. The press is a member of the Green Press Initiative, a nonprofit program dedicated to supporting publishers in their efforts to reduce their impacts on endangered forests, climate change, and forest-dependent communities.

The paper used in this publication meets the minimum requirements of the American National Standard for Information Sciences—Permanence of Paper for Printed Library Materials, ANSI 39.48-1992.

CIP data on file at the Library of Congress

26 25 24 | 5 4 3 2

Contents

In memory of *mi bisabuelo*, Dr. Aureliano Urrutia, a bridge between the arts and sciences, under which flow the textured waters of our heritage

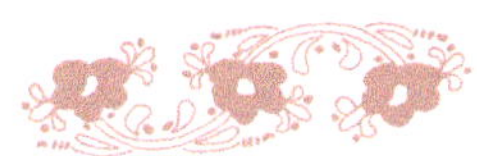

Árboles que no siembra el egoísta
Sino quien noble en el futuro piensa
Quien con desinterés, para el mañana
Deposita un legado de belleza
Y fervoroso, al porvenir confía
Útil y hermosa ofrenda . . .

—JOSÉ JUAN TABLADA

Foreword

TOMÁS YBARRA-FRAUSTO

The mid- to late 1940s, after the Second World War, was a period of optimism, hope, and shared aspirations throughout the country. In San Antonio, a popular verse for declamation went like this:

Como México no hay dos	There are no two places like Mexico
No hay dos en el mundo entero	Not two in the entire world
Ni sol que brille mejor	Nor even a sun that shines brighter
Como México no hay dos	There are no two places like Mexico

Those who recited the verse and those who heard it knew that indeed there are two Mexicos—one within the territorial borders of the Mexican nation and a second one, *el México de afuera*, comprised of the Mexican diaspora. *El México de afuera*—the Mexico outside of Mexico—includes Mexican Americans born in the United States who speak Spanish and/or English and retain the Mexican cultural heritage informed by their lived experiences in the United States, and the citizens of Mexico who relocated to the United States as exiles and migrants or immigrants.

Mexican nationals have crisscrossed the border since colonial times. Because of its historical ties and proximity to Mexico, San Antonio was a safe haven for exiles and immigrants after the turmoil of the Mexican Revolution. Those fleeing north included peasants, railroad workers, agricultural laborers, artisans, wealthy businesspeople, and entrepreneurs.

As a young boy, I recall seeing newspaper photographs of the elegant swells in their best attire who graced the social pages of the San Antonio newspapers. I knew about Dr. Aureliano Urrutia because he was my grandfather's *doctor de cabecera* (primary care doctor). My grandfather, Don Francisco Ybarra, was elderly but still handsome and dashing, with a robust white mustache and twinkling amber eyes. As his daily attire, he wore cowboy boots, well-pressed trousers, a Stetson hat, a starched white shirt, and a waistcoat with an elegant watch fob tucked into its pocket. Abuelito was always well groomed, and his distinctive appearance paid homage to the Spanish adage *La buena figura hasta la sepultura* (A positive presentation of self remains until the grave).

Abuelito had once been well to do. He owned a cotton ranch in Comal County near New Braunfels. Over time, he had gambled away his land, ranch, and possessions. Although impoverished in my youth, he still maintained the appearance of a proud *caballero*. He and my grandmother came to live with my family in our modest home on San Antonio's West Side. I was his only male grandson and became his caretaker.

On Saturdays, Abuelito and I took the bus to La Plaza del Zacate (Milam Park), where he was part of a group of elderly gentlemen who had coffee and conversed about current social conditions in the period of Jim Crow when Mexican Americans were segregated and subjugated. After the weekly session of political analysis, Abuelito and I would enjoy a taquito at one of the small cafes around El Mercado (Market Square) and take the bus home for a siesta.

From time to time, I would escort Abuelito to his medical checkups at the Clínica Urrutia at Houston and Laredo Streets. This flourishing downtown neighborhood was home to El Mercado, Mexican restaurants, the offices of *La Prensa*, and a Spanish-language bookstore that sold detective stories, cookbooks with *recetas caseras* (home recipes), *cancioneros* (songbooks) with lyrics of popular songs, books on Mexican history and culture, and literary works by San Antonio, Latin American, and European writers.

The clinic had an interior patio with a skylight and large *macetas* (pots) of *orejas de elefante* (elephant ears) and other tropical plants. A full-sized copy of Rogier van der Weyden's painting *The Descent from the Cross* and an elegant chandelier and stained-glass windows gave the waiting room a reverential atmosphere. The room was full of patients from all classes and walks of life.

A large leather sofa, comfortable chairs, and side tables with magazines to read as we waited were calming. When the nurse called my grandfather's name, we were ushered into Dr. Urrutia's inner sanctum—a consultation room with an examination table, an X-ray machine, and other medical devices. Framed religious paintings adorned the walls.

Dr. Urrutia would be sitting on a backless chair a few steps from the entrance. I would hold Abuelito's hand and carefully guide him to sit next to the doctor. Abuelito told me that Dr. Urrutia was a savant who could diagnose your malady on sight—by the time you walked over to him. He would explain his diagnosis and examine you thoroughly. At the end of the consultation he would give you other tests that generally confirmed his initial diagnosis.

His patients claimed that he combined the newest medical technologies with ancient knowledge of indigenous medicinal plants and healing practices. After his appointments, Abuelito and I would go to the Urrutia pharmacy and get the prescribed medicines before taking the bus home.

In spite of his fame and prodigious medical achievements, Dr. Urrutia remained humble, kind, and benevolent. He allowed patients to pay for his services according to their economic means. He died in 1975 at the age of 103.

Many anecdotes made Dr. Urrutia into a mysterious, flamboyant, extraordinary figure. He sometimes attended early mass at San Fernando Cathedral. Arriving in a chauffeur-driven limousine, he came through the front entrance in a dashing *capa española* (Spanish cape) that accented his indigenous features. Heads turned to watch him as he walked slowly to his marked pew, his huaraches making a swish, swish sound. The indigenous huaraches and the *capa española* were an incongruous combination.

In recent years, the City of San Antonio and Brackenridge Park have begun efforts to preserve Miraflores, a large garden created by Dr. Urrutia near what is now the intersection of Broadway and Hildebrand. Along with restoration of several art objects, a project has been initiated to restore garden paths and to plant grass and flowers. A venue for concerts, lectures, and other educational programs may be on the horizon.

The Miraflores sculpture garden has examples of *trabajo rústico*—or faux bois—benches and chairs, statues of Cuauhtémoc and Coyolxauhqui, antique Talavera reliefs, and other Mexican cultural and historical symbols. These objects are a veritable compendium of Mexican history and culture.

The beauty of Miraflores as a garden to delight and instruct calls out to be resurrected. Alongside the cacophony of a busy urban thoroughfare, Miraflores can provide a tranquil oasis, a place for contemplation. Visitors may once again amble down the pathways, sit and admire the plants and butterflies, and renew their connection with nature. As day becomes evening, the garden could once again become a quiet place to observe the Texan sky, its glimmering stars illuminating the serene landscape.

Dr. Aureliano Urrutia's garden has the potential to serve as a magnificent gift from the city—a reminder that *el México de afuera* continues as Mexican nationals migrate and immigrate to San Antonio. They refresh the language and expand the culture of Mexican Americans born in the United States. Miraflores, like the Spanish missions and the River Walk, gives San Antonio a unique cosmopolitan *ambiente*. The city is an American metropolis of many cultures. As it celebrates places like Miraflores, San Antonio envisions becoming a new beloved community that embraces the complexity of who we are and what we are becoming in a multicultural country that is more diverse, inclusive, and just.

Preface

How I Came to the Garden

For much of my young life my great-grandfather, Dr. Aureliano Urrutia, was a mysterious figure, even though he was mentioned almost everywhere I went in my hometown of San Antonio, Texas. It seemed that everyone else knew him, but I didn't. I was raised in silence about him, carefully guarded from and oblivious to the circumstances of his escape and exile from Mexico in 1914.

Throughout the years of my youth, as I visited the local downtown hospitals, accompanying my doctor father on his weekend rounds, the nurses on each floor gave me candy while he visited patients. "Your great-grandfather was an amazing doctor!" was their refrain.

I did not realize that he was alive until I met him toward the end of his life, in about 1974. He was more than a hundred years old, and I was thirteen. My father took me to visit. I remember the man as meek and quiet. His enveloping hands, glinting eyes, and kind smile met mine. We did not speak. The entire visit lasted no more than a few minutes, and the connection just seconds, but we made contact. Fortunately, I met him. I am sorry that I did not really know him.

That same year my father took me to Guadalajara to attend the wedding of his cousin, a twenty-ish granddaughter of Urrutia. There I met a vibrant, educated, creative, and celebratory family who showed me a new side of my identity. As I sat in a circle of perhaps thirty young people out on the lawn, a handsome young cousin next to me revealed that all were *primos*, cousins of some degree. In one moment my small San Antonio family expanded into an enormous *familia* in Mexico. The weekend evolved into a joyous celebration, with my young adult cousins whisking me through the most exciting nightlife of the big city. I also recall a midday feast at a huge long table in the open air, under a palapa. I fell in love with Mexico and cried profusely when we left, but I went home with a new part of my story.

The next summer, in 1975, my great-grandfather died. I saw the full-page newspaper reports and squirreled one away, the beginning of my collection and my first memory of reading about his life and Miraflores, the magical garden he created in San Antonio. Sadly my family no longer owned the garden, but soon after I wandered through its gates anyway.

Alone with my camera, the abandoned garden served as my first subject for a 1978 photography class taken during my final semester of high school. The wooded, overgrown landscape sat quiet and obscure, even ruined in places, as if no one had been there for decades. But I could tell that this had been a special place. I wondered why this great man's beautiful garden had dissolved into a tangle of weeds and crumbling decay. I took a roll of photographs, hand-developed the negatives, hand-printed the photographs, and added them to my collection.

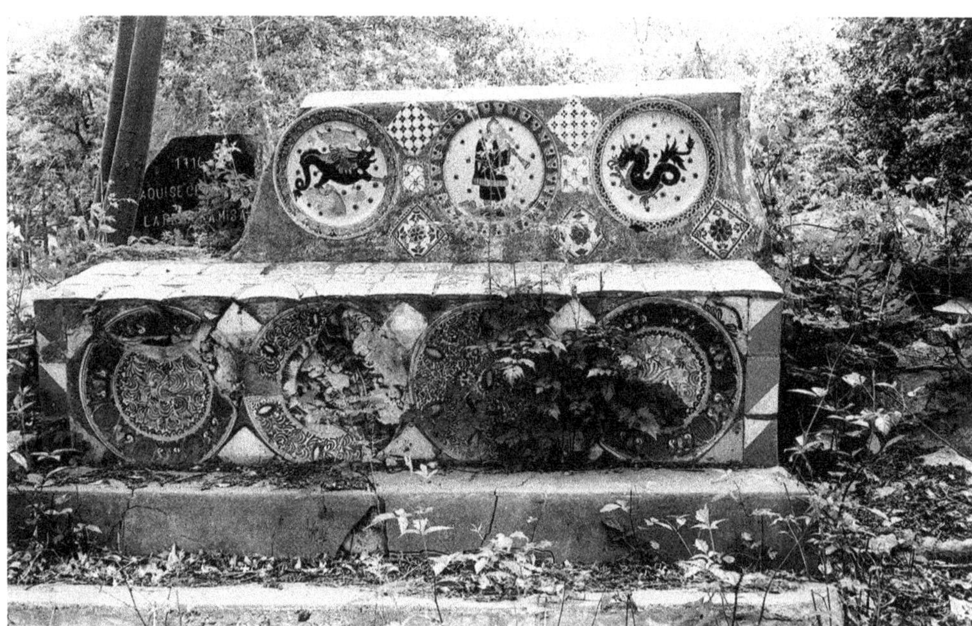

Two-sided bench, pond-side view, Talavera ceramics and concrete, in ruins.

In the years that followed, the story slowly got filled in, especially during some long drives with my father to and from Colorado College, where I attended school. Gradually, over many years, I learned more of who Urrutia had been—an indigenous man of Aztec descent, a youngster of prodigious intellectual abilities, a soldier married to a well-to-do woman of Spanish descent, and an accomplished surgeon. I also learned that this romantic picture collided head-on with the Mexican Revolution, forcing Urrutia into exile. He arrived in San Antonio in 1914 and settled his family here, where some still live today.

In 1982, only seven years after the death of Urrutia, his fourth son, Adolfo, my grandfather, died and passed along the doctor's archive to my father. My father was forty-seven; I was twenty-one. Over the next decades, new layers of detail emerged. Though highly accomplished and known as a physician and surgeon, Urrutia had been appointed to serve as secretary of the interior in the Mexican presidential cabinet, and his high governmental position subjected him to rumors of complicity in a hated administration known for its dictatorship and assassination of powerful opposition leaders. Life in politics quickly became complicated and dangerous, and the doctor found himself with no choice but to resign after three months in office and escape from Mexico.

Slowly, my father began to share what he had learned of his grandfather in an evolving lecture. He called his talk "The Other Doctor Urrutia." From these talks, I learned that in San Antonio, Urrutia had built a grand home on Broadway and a thriving medical practice that served tens of thousands of people from all walks of life. I also learned about his wife and children and their attributes and accomplishments. My father touched on Urrutia's struggle with issues of assimilation and exile, not always certain how to couch it. During Urrutia's life in San Antonio, unsavory rumors periodically surfaced and resurfaced, asserting his participation in political deaths that had occurred around the time of his government service and bringing recurring pain to him and his family, but lacking any credible or direct supporting evidence. On the other hand, Urrutia led an illustrious life in San Antonio, making important contributions to medicine and surgery worldwide, participating in civic life, and building and sharing his magnificent garden.

As I listened to my father, and to other conversations about Urrutia and Mexico's tumultuous history, I began to understand some of the underlying problems my family had experienced, even three generations removed, as a result of my great-grandfather's exile. I also began to understand the essential nature of the garden for Urrutia, and I helped my father to incorporate some information about the garden's landscape into his talk.

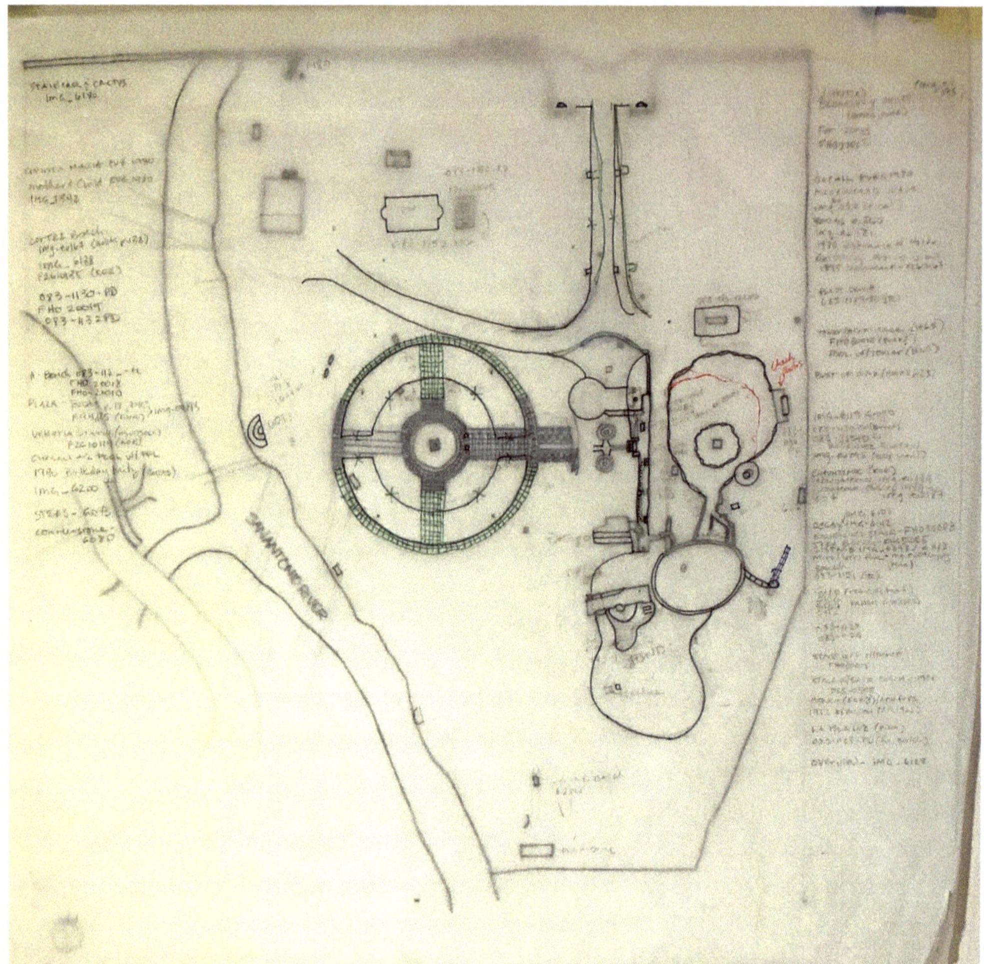

Author's initial sketch of Miraflores, 2017.

In 2012, as the hundredth-year anniversary of Urrutia's exile approached, conversations with my father intensified. He was seventy-eight; I was fifty-two. We often met on weekends to go through Urrutia's archive and discuss the details in his papers, photographs, and stories. And we made trips to the garden to forage for clues of the past, each time noticing new things, including the garden's rapidly declining state.

After a few years of meeting and talking, I expressed a desire to write about Urrutia, and my father supported that decision. Digging out my photographs and negatives and seeing that they survived thrilled me, as they would contribute to the record of the garden's condition over time. This book encompasses some of the ideas I've encountered, things I've learned, and conclusions I've reached about Urrutia, especially the significance of his garden at Miraflores.

About Miraflores

In Mexico, in the years leading up to Urrutia's exile, his remarkable success and charisma attracted other Mexico City thinkers in the fields of art, literature, science, and the other liberal arts. That interest in multidisciplinary thinking was characteristic of Mexican intellectuals of the time, and stayed with Urrutia throughout his life.

Urrutia was, like many others who were exiled from Mexico, intensely connected to his homeland. His challenge, then, became to rebuild something of the Mexico he knew and loved in his new home. The Mexico especially dear to him was Mexico City, a capital turn-of-the-century metropolis in the land of the Mexica with its deep history of the Aztec Empire, its complex layers of Spanish and European conquest, and its wide-reaching influence that spread with Mexican Independence throughout the country. Urrutia had a lifelong interest in plant life and gardens, scientifically and culturally, so it seems natural—albeit in hindsight, unusual and bold—that he would express himself by constructing a garden in San Antonio as an ode to his home country. Other personal gardens exist in the world, but Miraflores uniquely encapsulates Urrutia's Mexico, a multifaceted environment experienced not only by him, but shared to some degree by a diverse people over time and space. It gives us a window into so many aspects of that place and its history, with the remarkable perspective of seeing it via *el México de afuera*, in San Antonio, where Mexicans escaping the turmoil of the revolution joined an established Mexican American community. The perspectives of these refugees, however they are expressed, have great value, both to the Mexico that lost them and to the cultures that gained them.

Unfortunately, much of Miraflores no longer exists. Time, neglect, lack of understanding, and active destruction have combined to erase much of the landscape. We don't yet know how much of it will survive. So this book is written to create a memory of Dr. Urrutia's lost garden, and to explore the legends, symbols, stories, and ideas Miraflores holds for us today. It evokes a walk through the garden of Miraflores as it was during Urrutia's ownership of the landscape, from 1921 to 1962.

Although there were four to six entrances into the garden, the two primary grand entrances used by most visitors to Miraflores were the Urrutia arch, which is no longer present on the property, and the extant gate on Hildebrand, described by Urrutia as a "Monumento a la Ciudad de México," a monument to the founding of Mexico City. The discussion of the entrances begins with the Urrutia arch, but the walk envisioned in this volume begins at the Monumento a la Ciudad.

There are five chapters, one for each sector of the garden (entrance, esplanade, pond, plaza, and Quinta Maria). Each chapter ends with a sidebar focusing on an aspect of the garden (Talavera, artists, water, statues, and women) and highlighting some of the thematic threads that permeated Urrutia's thinking and expression.

The maps and diagrams reflect my research on and initial sketches of the original appearance of this now nearly destroyed garden. In the map of each sector, the features are numbered and keyed. The book also includes integral photographs with captions about various aspects of the garden.

Urrutia generated a number of writings and self-published volumes during his lifetime, including a small book of selections from his art collection in 1940.[1] His second daughter, Luz, managed the family archive until she passed away in 1945. His eldest daughter, Refugio, took on the responsibility of compiling Luz's collection into a family volume in 1946.[2] Excerpts from these and other contemporary writings reflect the understanding and observations of people who visited and wrote about Miraflores at the time.

Terms of Cultural Identity, Nationality, and Ethnicity

In recent years, there has been much discussion and confusion about how best to refer to people who originate from countries south of the United States. What do we mean when we use the word *Hispanic*? *Latino/a/x*? *Latin American*? Putting these terms into a historically based text complicates matters further. Add to that typical institutional and government uses of these terms, plus my own and other people's opinions on the subject, and reaching an agreement becomes impossible.

I use the following terminology in this book:

LATINO

A person with origins in or cultural ties to one or more of several Latin American countries. People from the originating countries may not refer to themselves as Latino (instead calling themselves by their country of national origin—for example, Cuban, Mexican, or Argentinian), but as a group or generally, they may be referenced as such.

HISPANIC

A person with Spanish or Portuguese colonial blood or heritage; the term may also refer to the period during or after conquest or colonization by Spain or Portugal. On a related note, because of the mixing of Spanish and indigenous populations in Mexico beginning in the sixteenth century, the population largely became *mestizo*, or mixed. The idea of *mestizaje*, or mixed heritage, carries major themes and values of a widely varied cultural identity in Mexico. The nuance of this mixing in Mexico's history is a highly complex topic in and of itself. The concept differs significantly from the history and development of the idea of cultural mixing in the United States, and this difference underlies some of the cultural differences—and perhaps some of the tensions—between Mexicans and some of their neighbors north of the border. In contrast, pre-Hispanic refers to a time before colonization or conquest, when the indigenous population of Mexico was free of such imposition or influence.

MEXICAN

A person born in or descended from Mexico; also refers to the country or its people. About 13 percent of Mexicans are purely indigenous, but the remaining population is largely Hispanic, being either of Spanish or *mestizo* heritage.

MEXICAN AMERICAN

A person originating or descended from Mexico who is assimilated into the United States or has US citizenship.

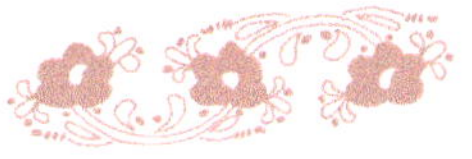

Introduction

Aureliano Urrutia was the creative mind behind his private garden, Miraflores, after the 1914 loss of Mexico as his home. The garden presented a view of Mexico from his eyes, as an exile of the Mexican Revolution (1910–1920). The mid-twentieth-century landscape, built chiefly from 1921 to 1945, portrayed the best of the Mexico he knew, from pre-Hispanic times until his exit. Miraflores was a work of art—a visual metaphor for Mexican history and tradition presented through landscape, architecture, and art. The garden sat near the source of the San Antonio River.

Urrutia was born in 1872 in Xochimilco, Mexico, a pre-Hispanic agricultural community that for centuries has grown food and flowers for the region using an ingenious water-based system of fertile human-made islands, or *chinampas*, built in a network of lake and canal systems. The town, known historically as a center of Aztec culture, is now a suburb of Mexico City and a World Heritage Site. Certain aspects of Miraflores were inspired by the unique water-based landscape of Xochimilco, visually referring visitors back to the place of Aztec origins where the Spanish first evangelized in Mexico. Urrutia's mother, Refugio Sandoval, most certainly of indigenous descent, passed away soon after his birth, and his father, Pedro Urrutia, perhaps ran a bakery or *molino* in the town.[3] Family lore has it that Urrutia's curious nature led him to explore his surroundings in great detail, inspiring and fueling his interest in many disciplines. Miraflores is one expression of his multidisciplinary approach to science, art, architecture, literature, music, performance, education, and history.

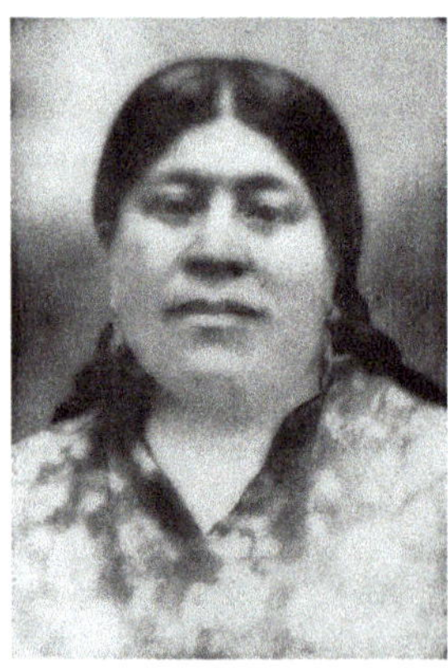

María del Refugio Sandoval Meléndez (1840–1873).

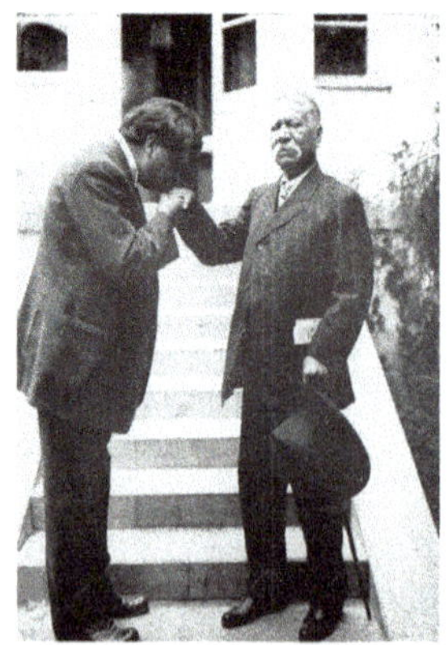

Aureliano Urrutia, left, with Pedro Urrutia Linares (1842–1918), ca. 1910.

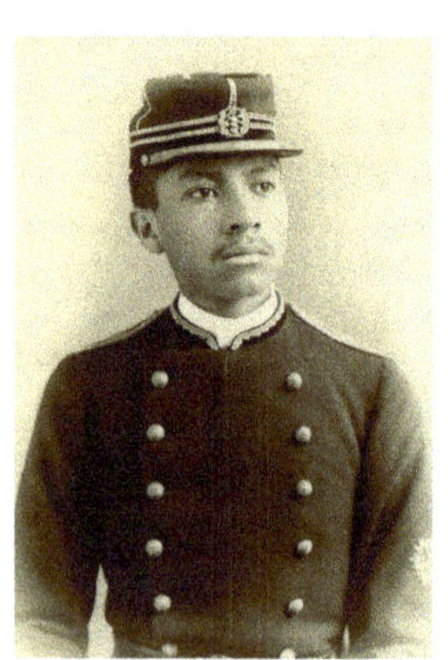

Aureliano Urrutia, 1892.

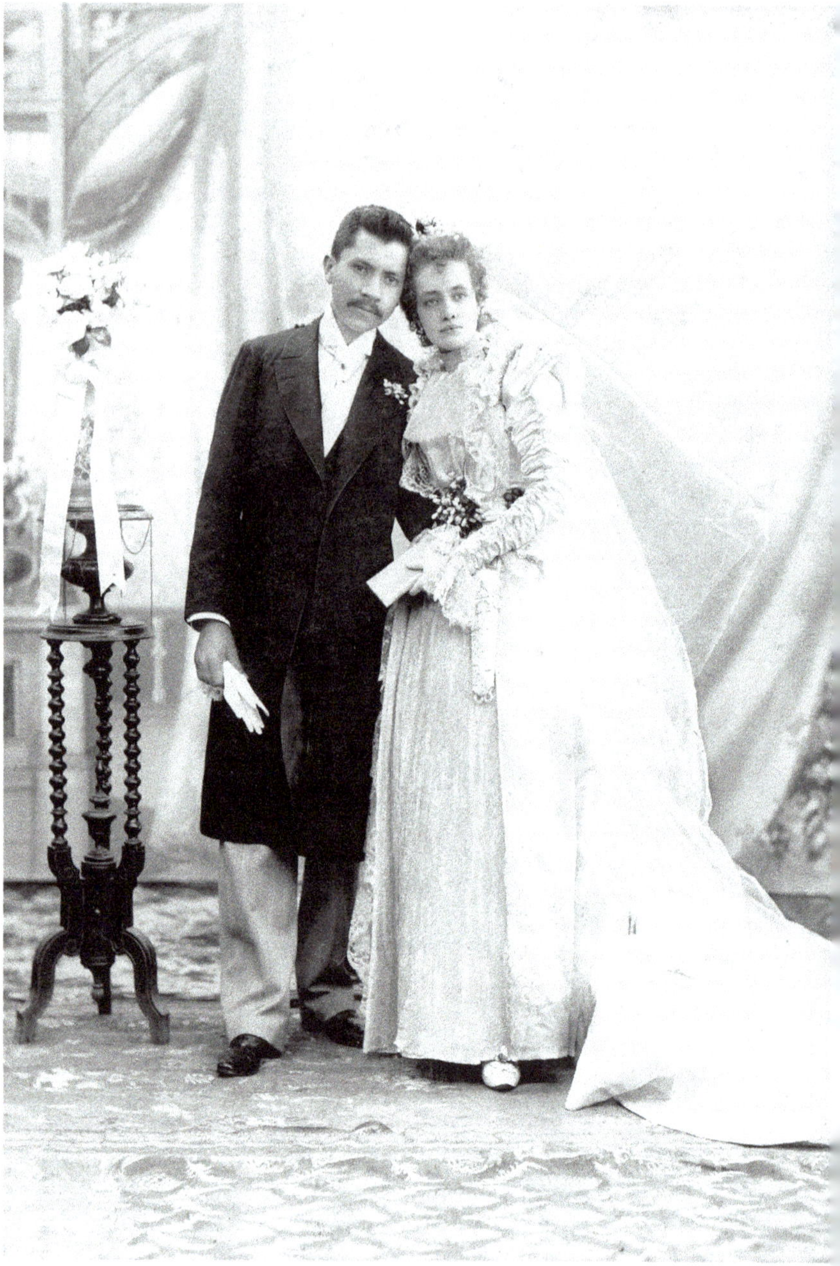

Aureliano Urrutia Sandoval and Luz Fernández Montáñez, 1897.

Although from a family of modest means, Urrutia was an ambitious child who grew to feel he owed a major debt to President Porfirio Díaz. Under Díaz's educational initiatives, Urrutia transferred from a local parochial school to a prestigious Mexico City high school. He excelled quickly and earned top national honors at his graduation. He met the president himself, and from him gained access to prestigious military service. With the military also came medical school training.[4] Díaz and the idea of education occupy a central place in the garden of this supremely educated physician. In 1897, he married Luz Fernández, the daughter of a businessman from Michoacán, and began a two-year stint in the medical service for the Mexican army, settling into private practice in 1899.[5]

The marriage to Luz, a highly valued union of indigenous and Spanish, over the next fifteen years produced twelve children: eight girls and four boys. Luz certainly inspired Urrutia's creation of Miraflores, the property for which was purchased shortly after her death. An atmosphere of feminine influence permeates the garden, including an area specifically dedicated to her memory.

Miraflores was not Urrutia's first garden, and San Antonio was not his first successful medical practice. At the turn of the century, as Mexico strove to emerge onto the world stage in science and industry, Urrutia's research and successful practice in the fledgling field of modern surgery gained him recognition as one of Mexico's top surgeons. By 1900, at the age of twenty-eight, Urrutia was appointed as head professor of surgery to the Escuela Nacional de Medicina de México.[6] He was its youngest professor.

By 1911, he had built a celebrated hospital complex, Sanatorio Urrutia, in Mexico City's historic suburb of Coyoacán.[7] Designed in the burgeoning European style of the time, the twenty-five-acre compound housed a state-of-the-art medical facility set on pastoral grounds with thematic gardens, sculpture, and paths for strolling, contemplation, and recovery.[8] Its opening was celebrated and attended by Mexico's president, Francisco León de la Barra, along with future president Francisco Madero, their large families, and other prominent dignitaries.[9]

Ultimately, Sanatorio Urrutia also came to have a reputation as a center for teaching in the field of medicine and as a salon for intellectual discussion. Urrutia's circle of friends and colleagues—influential writers, historians, musicians, lawyers, artists, and doctors—congregated to explore the most intellectual and engaging topics of the day. The discussion stretched far beyond the reaches of science and reflected a philosophy of interconnected fabric—that is, the idea that in order to be healthy, people must be nurtured by many disciplines. Miraflores, albeit on a smaller scale, also reflected the multilayered nature of Urrutia's interests and thinking, spanning time, place, and disciplines.

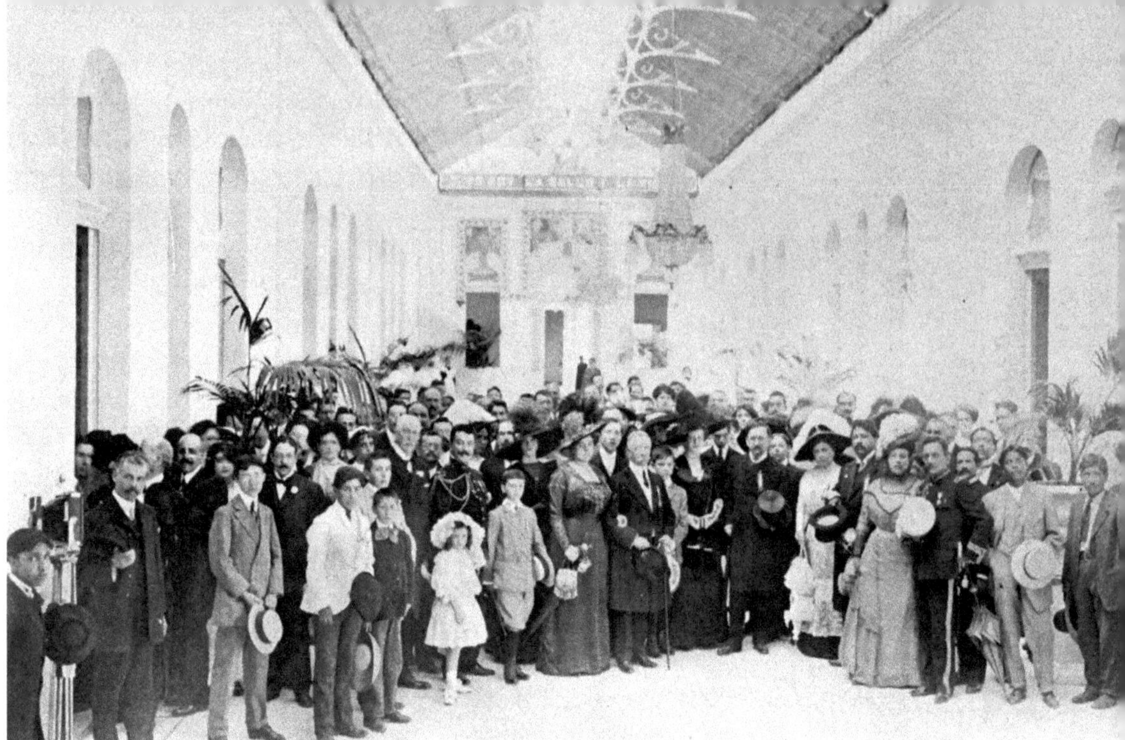

Grand Hall,
Sanatorio Urrutia, 1911.

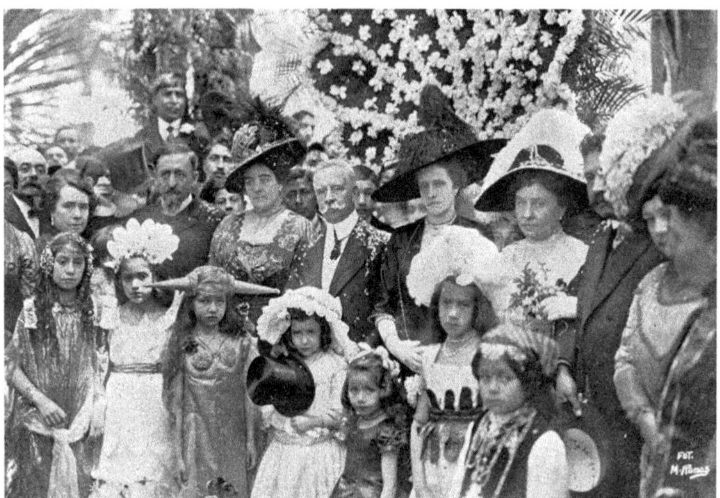

Opening festivities at Sanatorio Urrutia, 1911. Francisco Madero, left, President Francisco León de la Barra, center, with Luz Fernández Urrutia in black hat, center right. Urrutia stands partially obscured at far right.

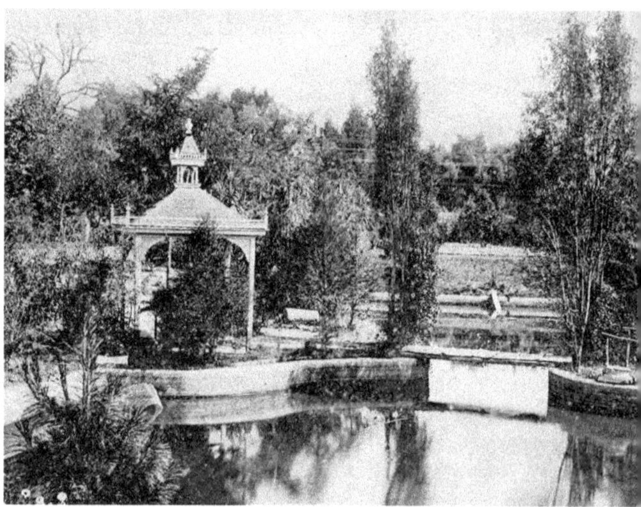

Recuerdo de Xochimilco, a garden at Sanatorio Urrutia, ca. 1911.

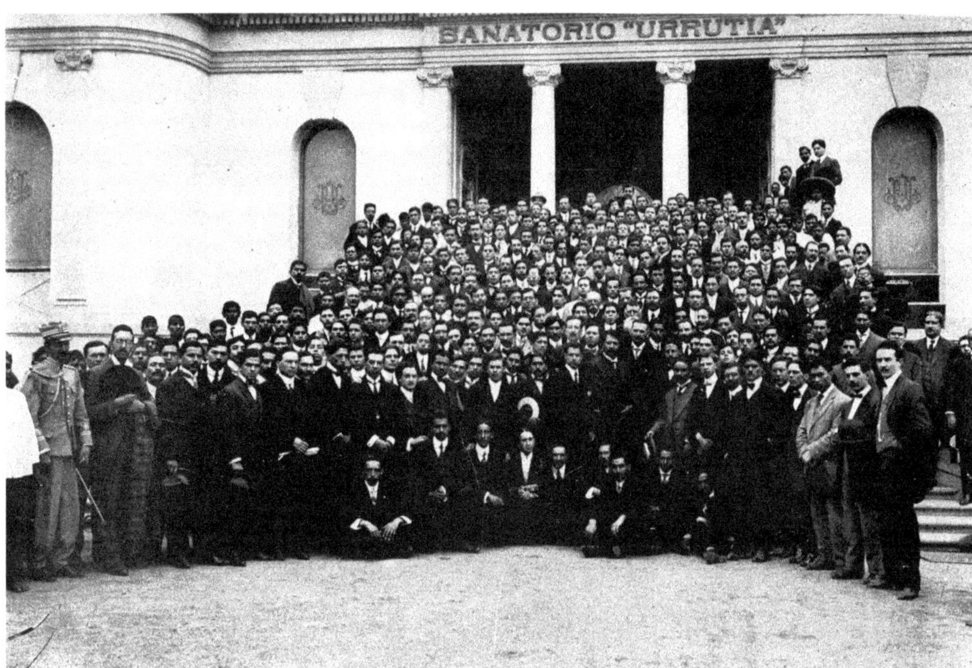

Supporters of Urrutia in his government appointment, ca. 1913.

In 1912 Urrutia rose to director of the Escuela Nacional de Medicina de México, where he increased his following as a revered teacher.[10] Unfortunately, the idyllic setting of Coyoacán belied the impending chaos and anarchy of the Mexican Revolution. As a young surgeon, Urrutia had first caught the attention of Victoriano Huerta, then a young colonel, serving as his medic in an army battalion. The doctor saved the life of a soldier by performing a dramatic tracheostomy after a battle.[11] Ironically, this feat endeared Urrutia to the rising general, who over the ensuing eighteen years would engage Urrutia as his physician and confidant, setting the doctor on the path toward exile. In summer 1913, Huerta appointed Urrutia to his presidential cabinet as the secretary of the interior. Ultimately, Urrutia resigned after less than one hundred days in office, disillusioned and in an effort to disassociate himself from Huerta's disastrous political activities.[12]

For a short time Urrutia resumed his directorship at the school of medicine and accepted the directorship of the Hospital General.[13] But eight months later, in May 1914, he fled, permanently leaving his Mexican life behind under threats of death from Huerta himself. The Urrutias and their six eldest children narrowly escaped Mexico City, pursued also by factions hostile to Huerta, and were escorted from the country under the

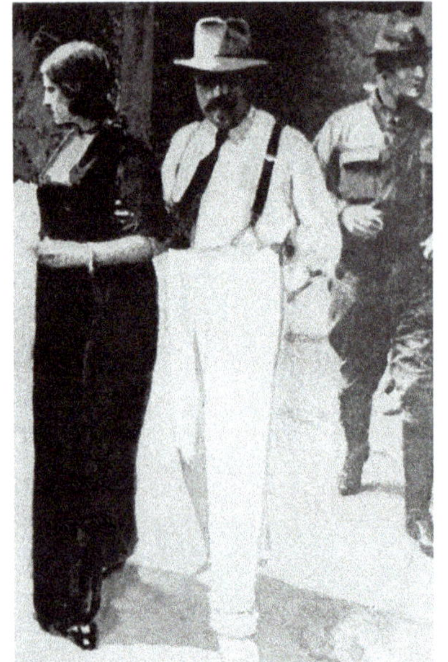

Urrutia and Luz Fernández, presumably at Veracruz, 1914.

Ship manifest, May 23, 1914. Urrutia was granted diplomatic status by the United States for his family's transport as the sole passengers aboard a military ship from Veracruz to Galveston.

armed guard of US troops at Veracruz.[14] Five younger children, secretly left behind, were brought out soon after.[15]

Of this period, Urrutia's words are few, but over the ensuing decades his friends, many of them illustrious Mexicans of the time, visited or wrote him in his new homeland, consoling him with their letters, poetry, and friendship. Some of their writings marveled about Miraflores, and even the Mexican press was fascinated by Urrutia's garden and the life he built in San Antonio.

Life in San Antonio: Success in Exile

After immigrating to the United States in 1914, Urrutia dedicated himself to building a life in San Antonio where he could raise his family and continue his medical career. Before purchasing the land for Miraflores, he set up two other properties: 420 Avenue C, which

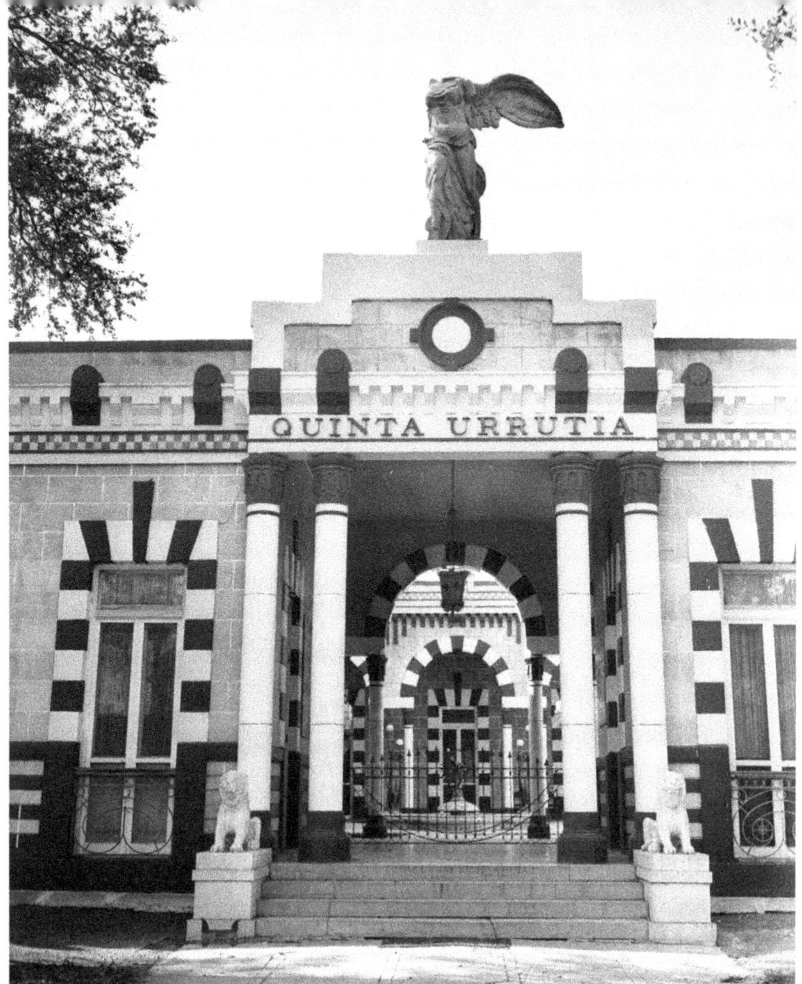

Front entrance to Quinta Urrutia. Nonextant.

became his office, and 3225 River Avenue, which became his home, Quinta Urrutia.[16] Both dusty roads soon thereafter became known as Broadway, a major north-south thoroughfare from downtown to San Antonio's prosperous north side.

The home, a ten-thousand-square-foot Spanish-style mansion designed by a lifelong friend, Porfirio Treviño of Monterrey, Mexico, sat among five acres of gardens with fountains, benches, and sculpture—a hint of the garden to come at Miraflores. The residence housed Urrutia's entire family of thirteen, a nanny, a cook, a butler, a driver, and several others in his employ. Urrutia's wife, Señora Doña Luz Fernández de Urrutia, ran the house adeptly and was recognized as "one of the best dressed women of the Southwest."[17]

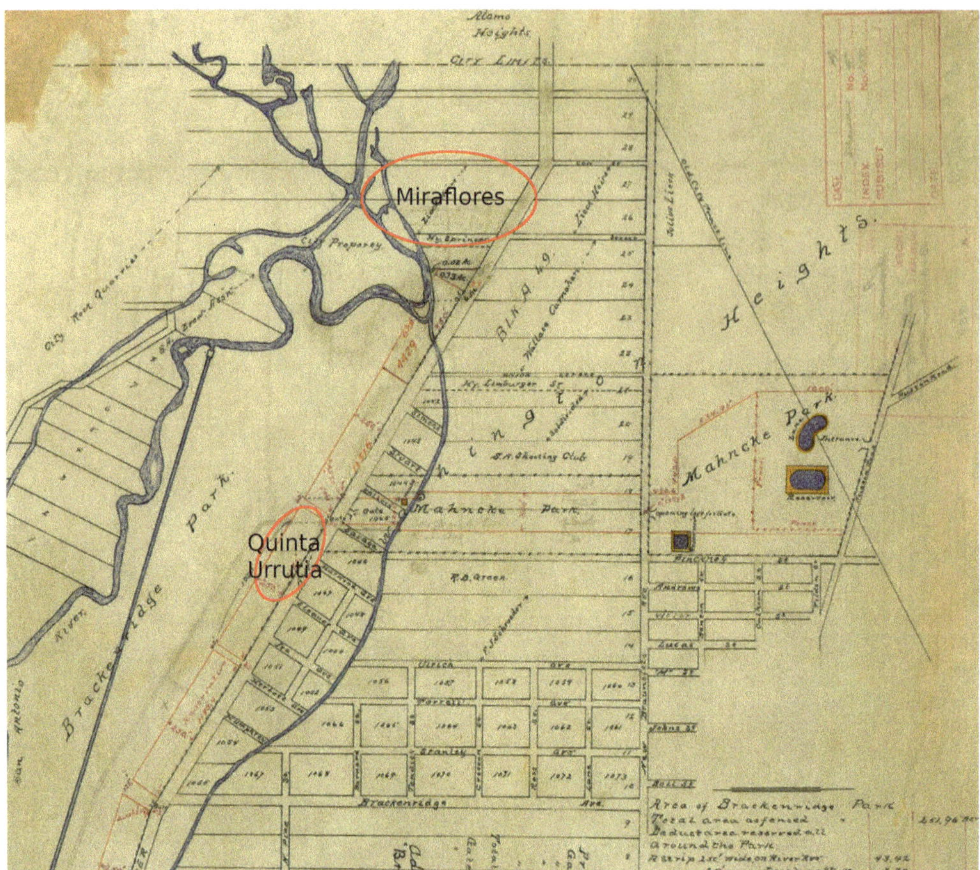

Miraflores's proximity to Quinta Urrutia.

Miraflores: A View of Flowers

Shortly after Luz's death in 1921, Urrutia purchased the land—part of San Antonio's original land parcels established by its Spanish mayor in 1731—for Miraflores from short-term owners from Mexico City, Margarita and Guillermo Alonso.[18]

Although Urrutia was highly successful in continuing his medical career, and thankful for the openness of San Antonio and the United States, he profoundly experienced the separation from his beloved home country since his immigration at age forty-two. Like many exiles from Mexico as a result of the revolution, he needed a way to stay connected to his homeland.

Exiles who could not return often expressed themselves by writing memoirs, handing down family traditions and values, and participating in community or political action.[19] Urrutia also did these things, but his passion for the land motivated him to express his *mexicanísimo* by creating an intricate garden. No other exile expressed themselves in the creation of a physical place that referenced their previous life in Mexico through art, artisanship, and landscape.

By the end of 1921, at least one artist had finished a sculpture for the garden, and a tile gate on Hildebrand had been built. By 1923, a small country house, Quinta Maria, was present on the property, and in 1924 Urrutia invited parks commissioner Ray Lambert to visit and see the garden's beautiful gates.

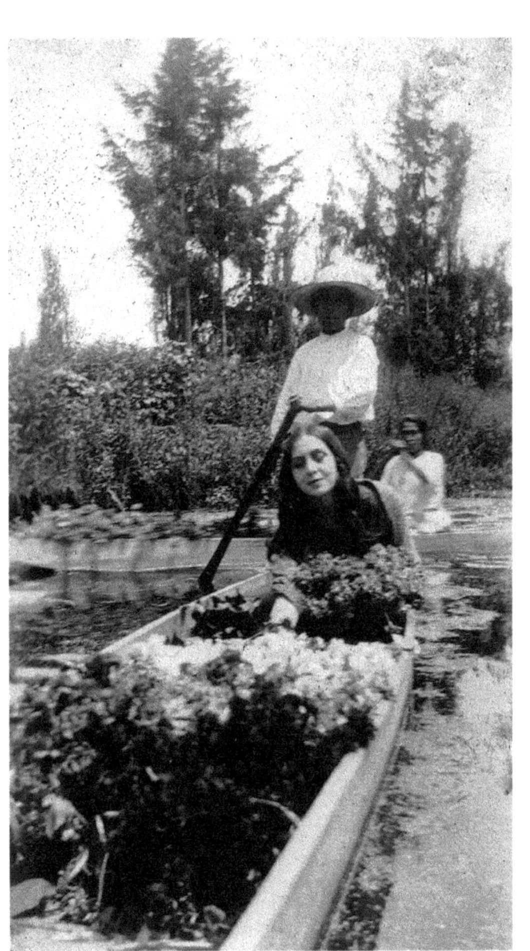

Urrutia's and Luz's daughter Emma, Xochimilco.

Over the next several decades, Urrutia continued to design and edit the landscape, and he and his children entertained friends, dignitaries, journalists, and more from the United States, Mexico, and other countries. Everything in Miraflores—the plants, architecture, sculpture, and artisanship—formed an atmospheric landscape reflecting Urrutia's connection to and understanding of Mexico. He commissioned Mexican artists and artisans to create sculptures and fountains evoking Mexican ideals. He planted species native to Mexico. As a result of his efforts, during its lifetime, Miraflores was the subject of many poems, articles, and photographs.

Even the name of the garden is an echo of the translated meaning of Urrutia's birthplace, for Xochimilco is a Nahuatl word that translates as "garden of flowers." As for *Miraflores*, the name is a stylized and interesting word of Spanish origin. Here again the merging of indigenous and Spanish is valued and exemplified. The word refers to, among other things, Rogier van der Weyden's highly symbolic fifteenth-century painting named for a Spanish monastery, the sixteenth-century flowered seaside area of Lima, Peru, and a set of locks on the Panama Canal. It does not have a direct English translation, but suggests a "flowered view." The word *mira* also translates as "watchtower," an architectural structure of which Urrutia happened to be fond.

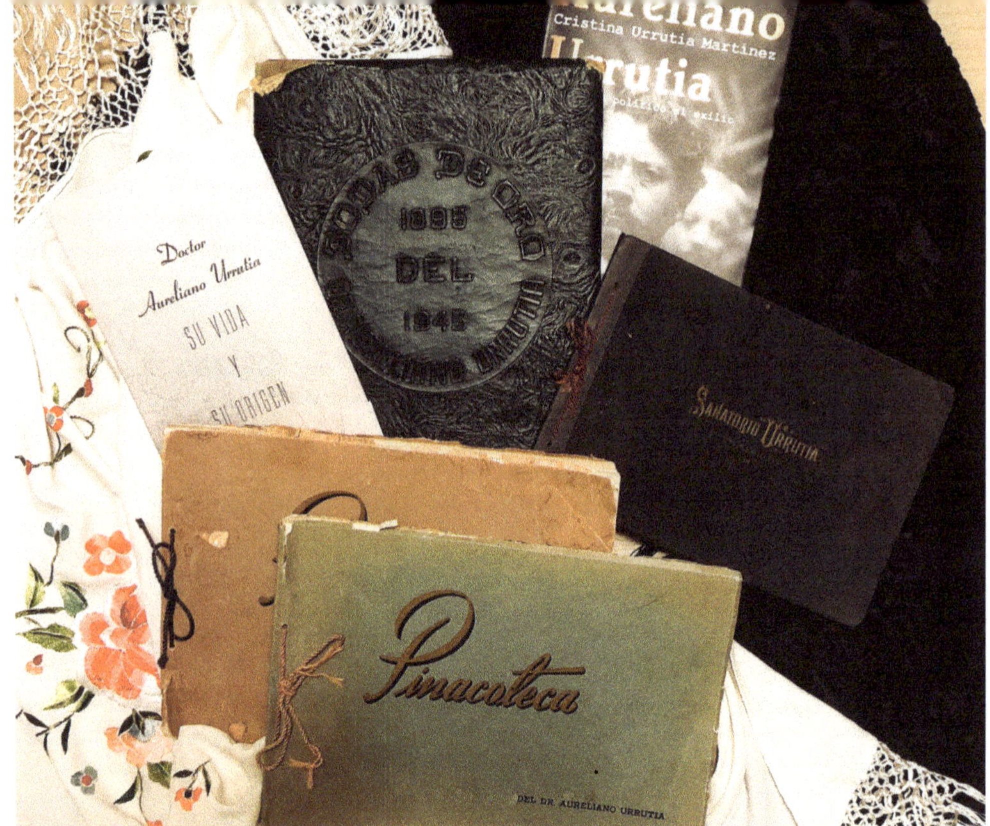

Self-published family volumes; with Cristina Urrutia Martínez's biography of Urrutia, published in Mexico.

Urrutia: The Artist-Scientist

Urrutia's friends often referred to him as a bridge between the arts and the sciences. In his Mexico salon were said to meet "*la flor y nata del arte y de la ciencia*" (the cream of art and science).[20] The famed writer José Juan Tablada, Urrutia's friend, wrote to him from New York in December 1927 reminiscing about their former lives in Mexico: "the flowery sanatorium—citadel of Coyoacán where you developed harmoniously your admirable work of science, of beauty. There the blood of the wounds you cured turned into the roses of those flowers in which you cultivated your most delicate and poetic feelings about Xochimilco."[21]

In Urrutia's interconnected world, whether in Mexico or the United States, he consistently interacted with people, from his tens of thousands of patients and the students and disciples he taught in the surgical theater to the visitors to his gardens, including Miraflores. Art, architecture, science, literature, history, folklore, culture, religion, plants, people, and places were all bound together as one in Urrutia's mind. Miraflores encompassed them in the beauty of a garden. It would not be lost on Urrutia, as a lover of the arts, that Miraflores would further survive in a book. In the following pages, we enter the garden, exploring the landscape as it appeared during Urrutia's ownership.

NOTE: Mexico has a brilliant tradition, when it restores its ancient archeological sites, of inserting small pebbles into the mortar that holds together the stones of pyramids, walkways, and other built features. I often wish for that sort of transparency in all restoration. In this book, the current condition of various objects in the garden is indicated in the photo captions describing any particular object. I have classified objects as nonextant, in ruins, damaged, restored, and replaced. Also, any feature that is no longer in the garden (that is, »nonextant«) is loosely identified with »reversed double angle brackets, also called reversed guillemets«. I've tried to use this convention for landscape features and art objects to give readers a visual sense of how much of the original plantings and hardscape have been erased, although even this approach fails to do them justice.

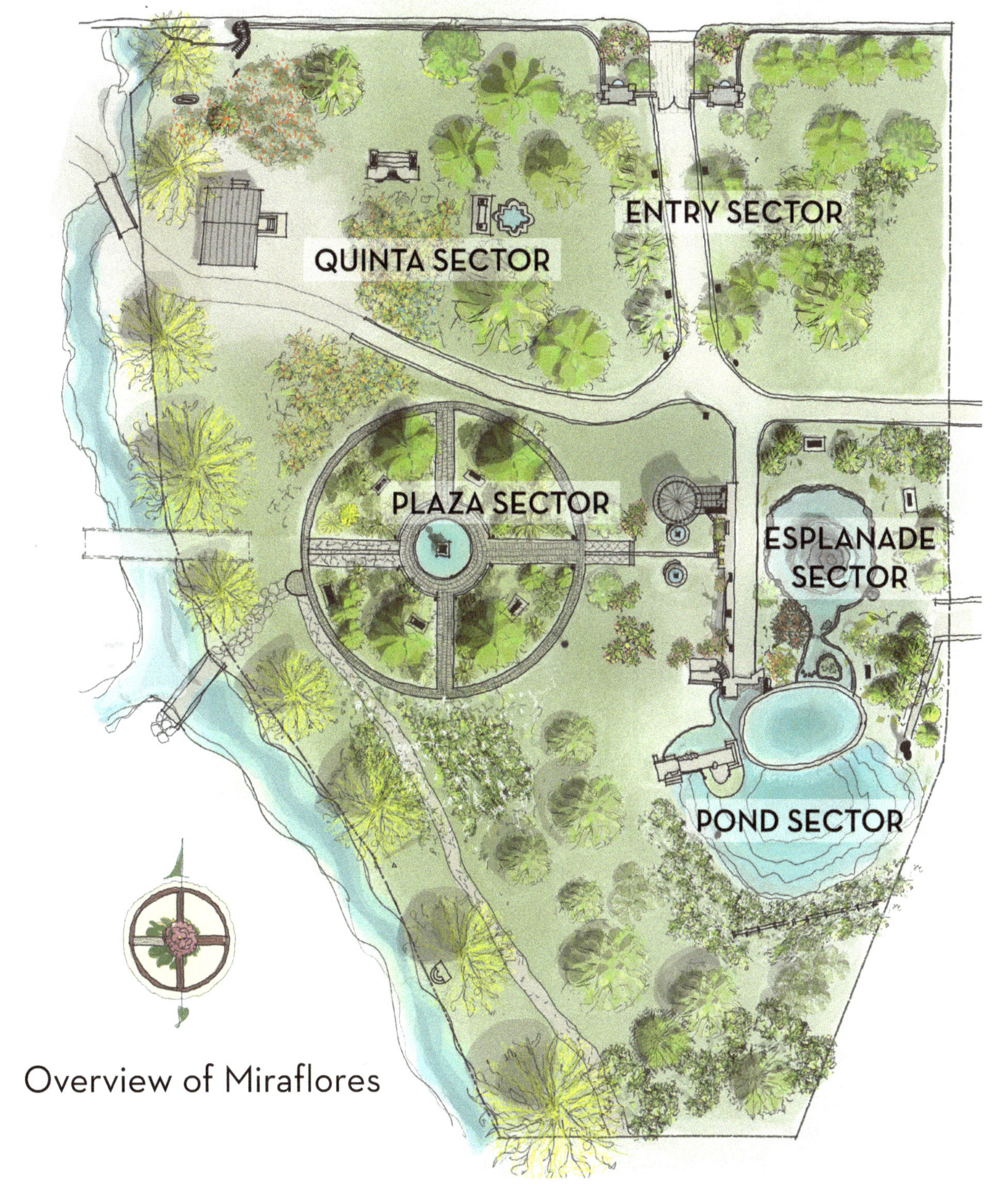

Overview of Miraflores

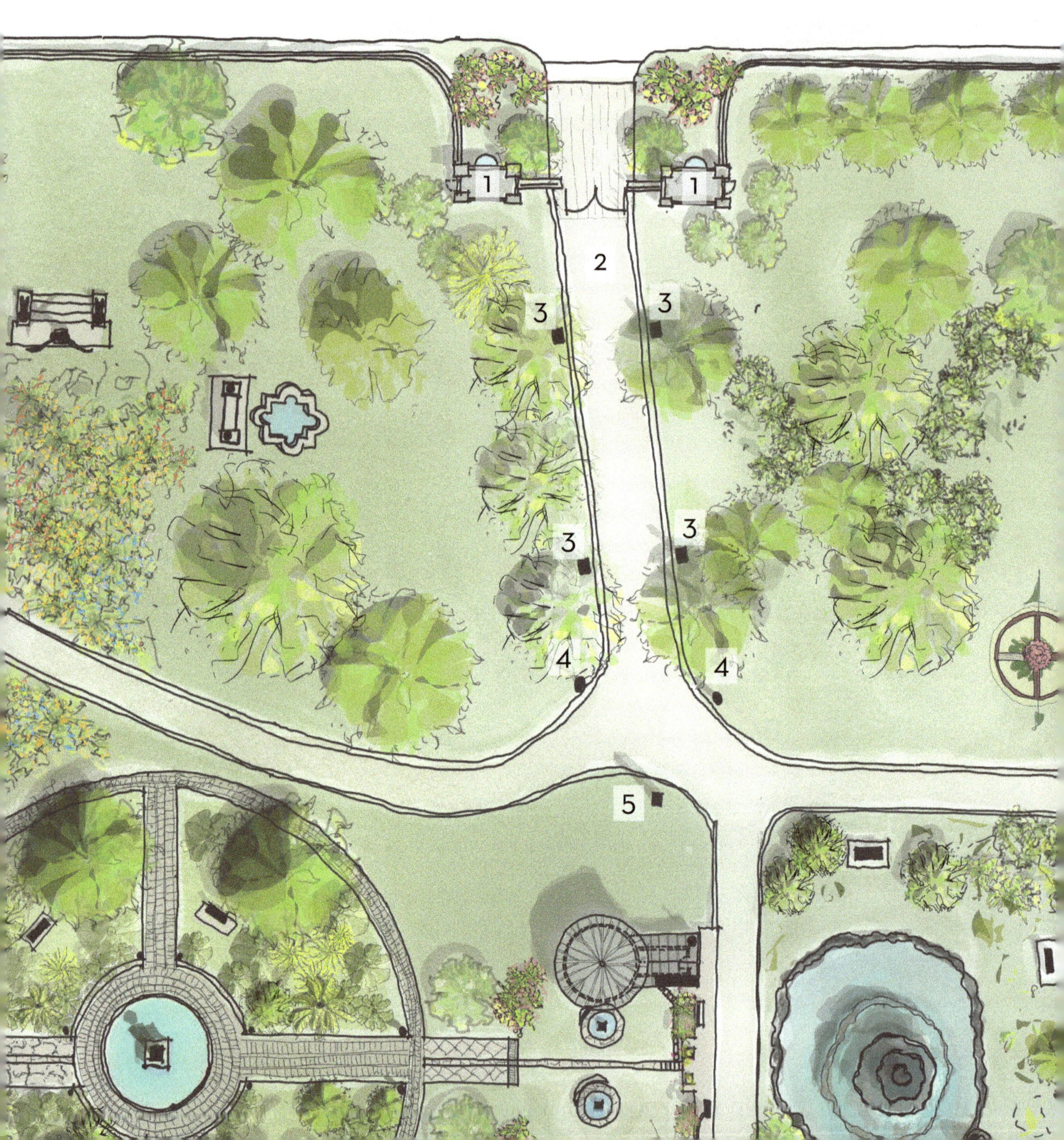
1
1
2
3
3
3
3
4
4
5

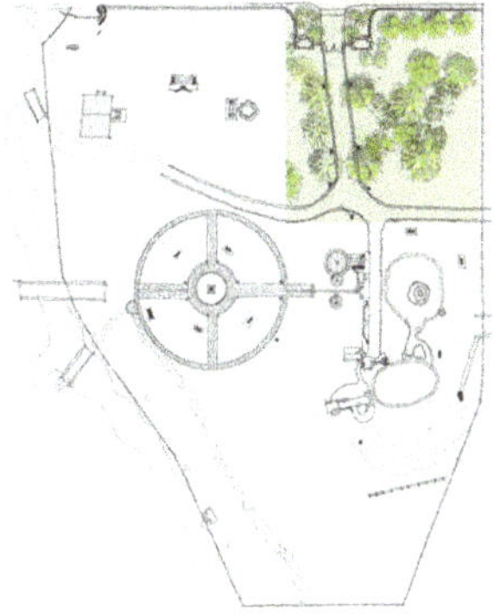

Entering the Garden

To walk into Urrutia's Miraflores at the height of its creation is to walk into a different world. The Talavera-tiled gates—the Urrutia arch on Broadway and the Monumento a la Ciudad de México on Hildebrand—startle. The process for making Talavera originated in Spain and was brought to Mexico after the conquest. By the seventeenth century, the artisanship of indigenous Mexicans became a highly regulated industry of ceramics characterized by a tin glaze over clay from specific regions. Both entrances at Miraflores are quite public, situated along prominent roadways, Broadway and Hildebrand, where people walk and travel by vehicle. San Antonio has never seen the likes of such colorful Mexican tile as this. Much like the classic fairy tale tent in Scheherazade's story of Prince Ahmed, these gates portend another reality, an adventure to an unfamiliar place.

- * Urrutia arch (see illustration at right)
- 1 Monumento a la Ciudad de México
- 2 Calzada del 2 de Abril
- 3 Commemorative columns
- 4 Taza pedestals
- 5 Monumento del General Porfirio Díaz

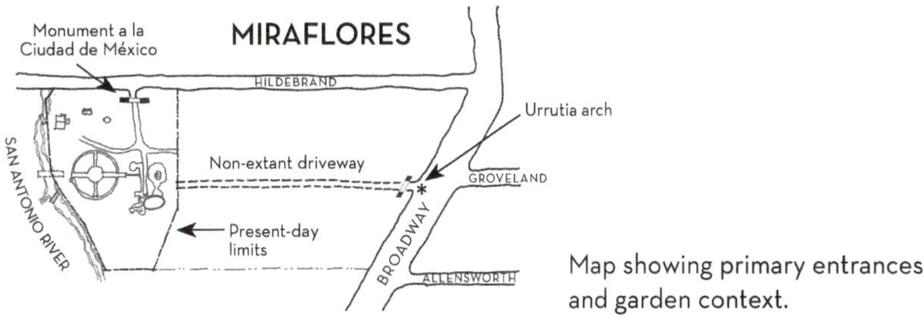

Map showing primary entrances and garden context.

Monumento a la Ciudad de México. Each tower is 18 x 14 x 9 feet, joined by a 40-foot wrought iron fence and gate, restored and some murals replaced.

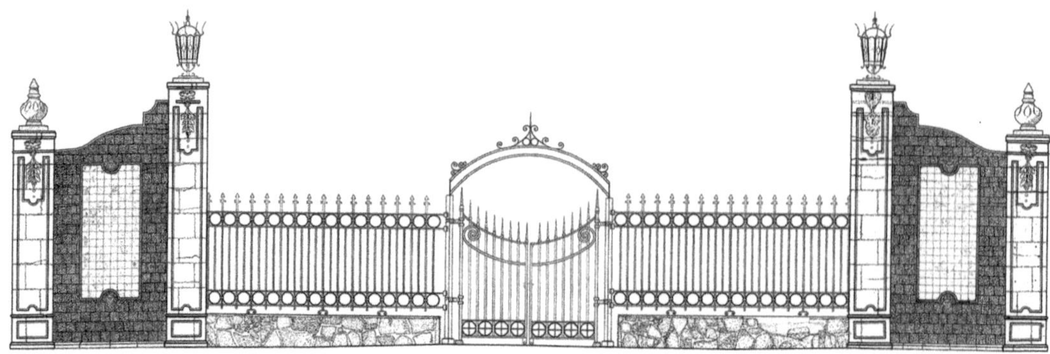

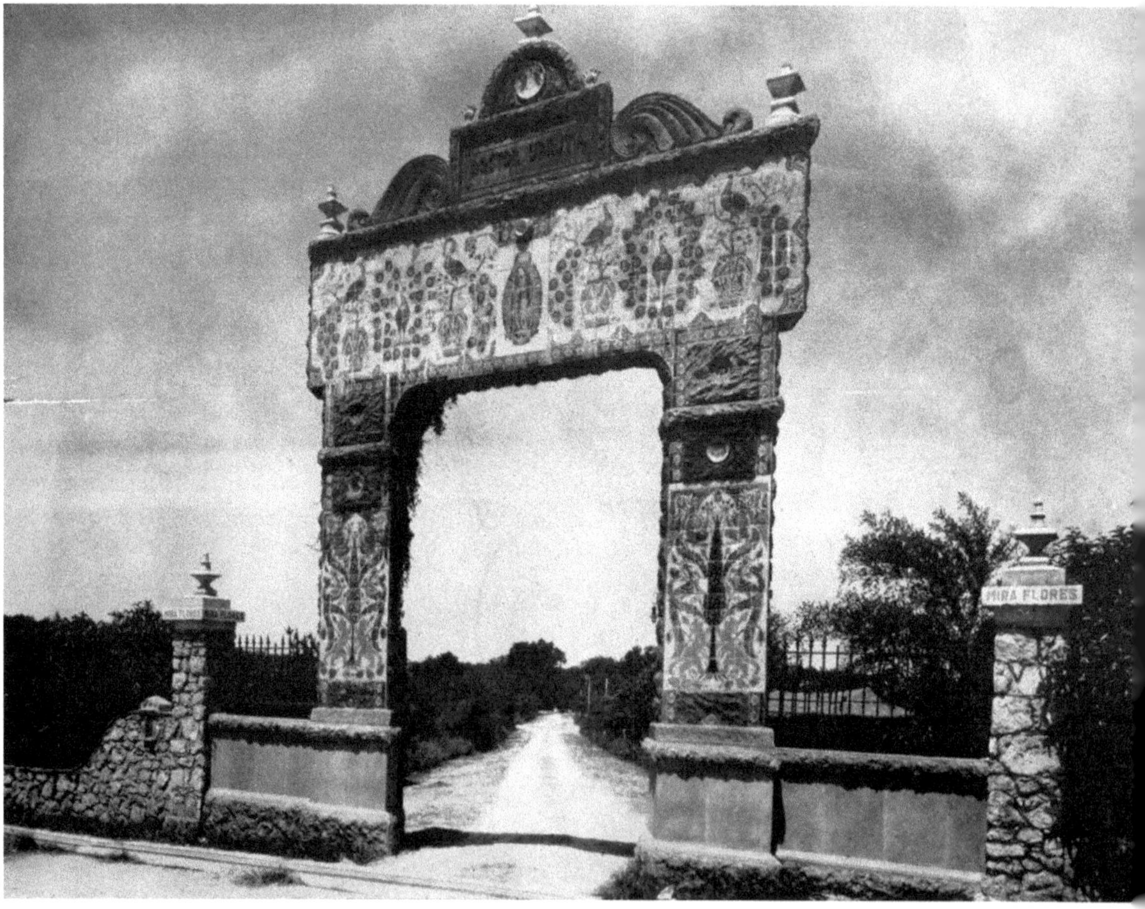

Urrutia arch, 24 x 20 x 3 feet, now in disrepair.

Outside the gates of the fifteen-acre property the day goes on as usual. Broadway (formerly known as River Road, and as Avenue C closer to downtown) shuttles traffic along its eastern boundary, including streetcars on a short-lived line, from downtown to these northern outskirts. Along the garden's western boundary the San Antonio River meanders just downstream from its bubbling source, the Blue Hole and other nearby springs. The stately Victorian-era Brackenridge Villa to the north has sheltered the Sisters of Charity of the Incarnate Word since 1897, alongside the 1907 redbrick, peaked Motherhouse Chapel of the Incarnate Word College. The forested Brackenridge Park completes the view to the west across the river and to the south, all the way down to Josephine Street.[22]

The Urrutia Arch: Bridging Worlds

The Urrutia arch rises two stories above Broadway, one block south of Hildebrand Avenue, across from Groveland Place. Ornate wrought-iron gates and designs using Talavera tile on an architectural scale were familiar to Urrutia, who saw them daily when he lived in Mexico. But here may stand some of the earliest Talavera tile used decoratively for a vertical exterior surface in the United States.

This primary entrance to Miraflores incorporates a variety of materials—Talavera tile and pottery, faux volcanic stone, concrete, and metal. The fanciful signatures on the tile indicate that it originates from the Uriarte tile workshop of Puebla, Mexico.[23] Mexican artist Dionicio Rodríguez contributed his faux stonework to the archway—perhaps his first commission in the United States. And a young artist, Luis L. Sanchez, created the concrete forms at the top of the lintel. Urrutia will go on to commission Mexican artists exclusively for all of the artwork at Miraflores.

At the top of the arch a ceramic plate displays a regal crest bordered by two birds of prey facing each other, wings spread. The crest rests over the bronzed words "Doctor Urrutia," all contained under an arch flanked by two refined, wing-like concrete scrolls. The collection of cultural symbols below reflects Urrutia's interest in various influences—royal, religious, natural, and Aztec—all somehow balancing each other.

Front and center, the lintel features the Virgin of Guadalupe in one of the city's earliest permanent public displays of her outside of a church environment. One of Mexico's most venerated images, the original depiction of the Virgin (1531) in Mexico City was officially recognized and celebrated by the pope in 1895 as Urrutia was emerging as a young

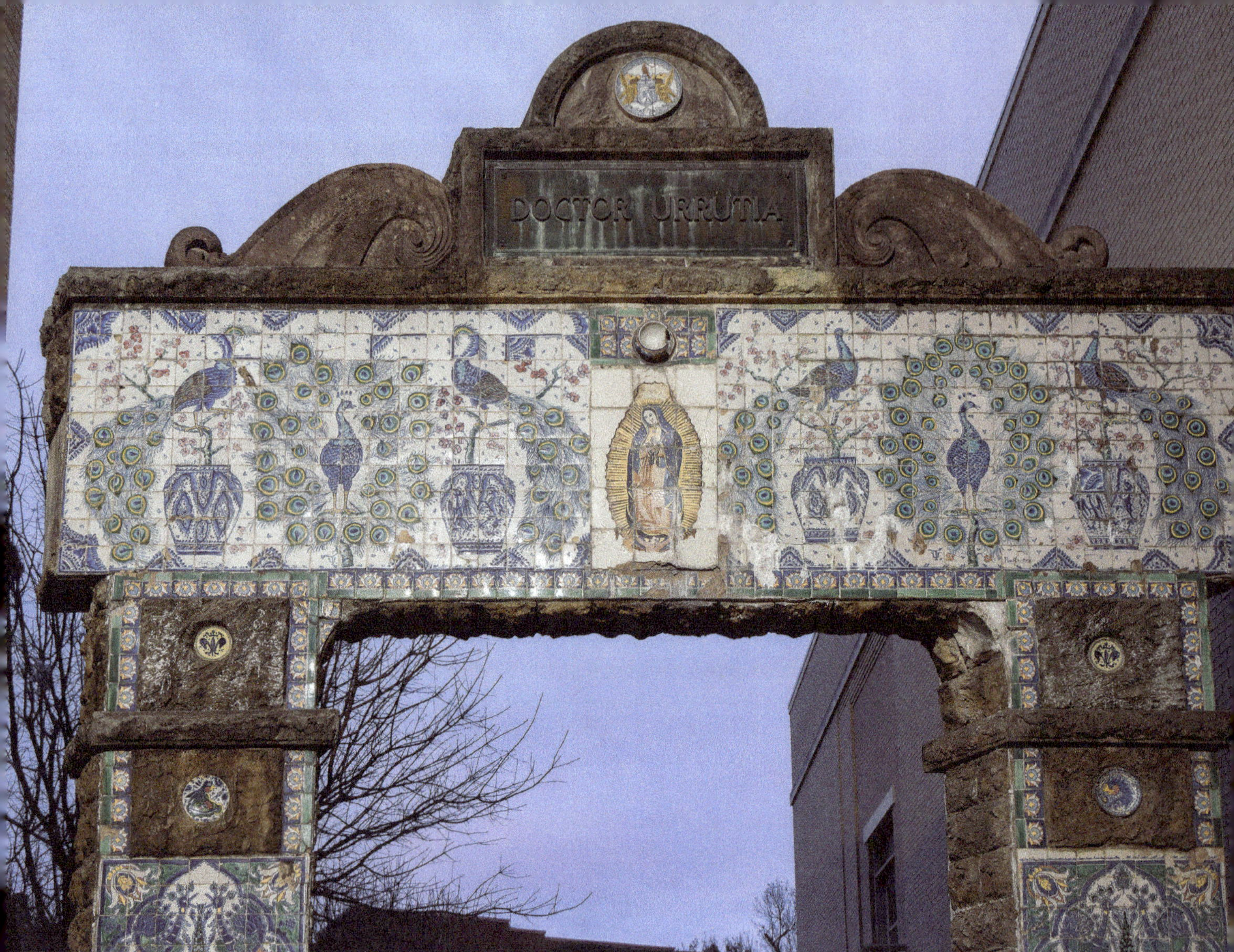

The Urrutia arch was acquired by the San Antonio Museum of Art in 1997.

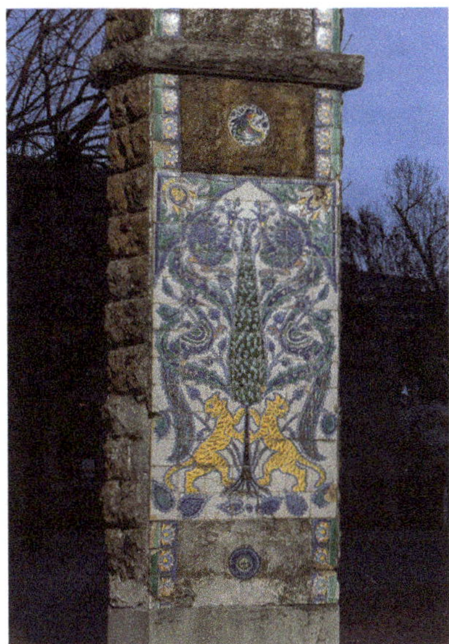

Urrutia arch column, detail.

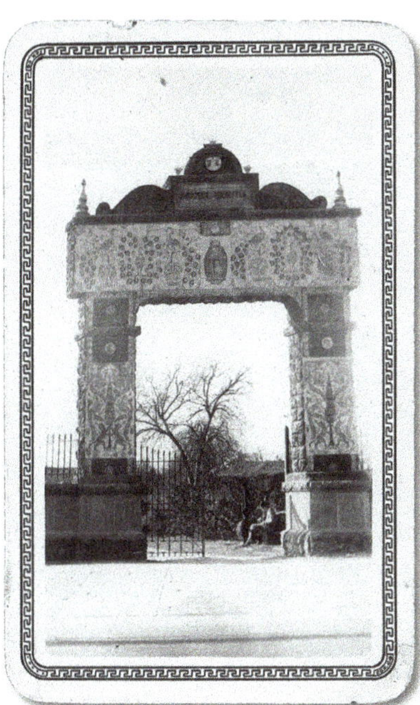

Urrutia arch.

surgeon. The story of the 1531 apparition of Our Lady of Guadalupe to the humble, indigenous Juan Diego goes to the very fabric of Mexican history. Her image was featured on both Father Miguel Hidalgo's 1810 flag of Mexican independence and Emiliano Zapata's 1910 flag of the Mexican Revolution.

Flora and fauna abound as six peacocks roost in blossoming cherry trees surrounding the image of the graceful woman. On each column a juniper tree—a tall variety of cypress native to Xochimilco—hosts more peacocks, surrounded on each side by rampant jaguars.

Peacocks are understood across many folk cultures to portray the fine line between beauty and vanity or to represent the notion of "all-seeing eyes" in their tail feathers. The jaguar, on the other hand, in Aztec tradition, carries a distinctive message of protection and warriorship not universally common, thus easily misunderstood. The cultivated cherries and sturdy junipers herald a perpetual spring for the animal and spiritual inhabitants thriving on the arch's columns.

The scene is upheld by volcanic stone, brilliantly fashioned to include the viewer's reality by mimicking the angle of the intersection of Broadway and Hildebrand, reinforced by the direction of the entrance road. It's a crossroads. In Urrutia's vision, disparate forces can exist together without conflict; here Mexican cultural icons join as sentinels of the garden, heralds of a bygone world. Before such an imposing entrance, a passerby might desperately wonder what lies beyond the gate but would not dare venture in unless invited.

From the Urrutia arch, a theme central to the doctor emerges: beauty comes from many places, and harmony results from coexistence. Most importantly, the archway communicates Urrutia's domain as rich, diverse, sophisticated, and strong. His portrayal of this complex Mexican expression, involving Mexican artists and Mexican materials, offers a glimpse of Mexican culture. In this gate, the arts, science, religion, and indigenous culture combine to welcome any friendly visitor to Miraflores. The long dusty driveway traverses acres of thorny mesquite brush into a Mexican garden—a wooded landscape with gravel roads, concrete and brick walkways, and winding footpaths punctuated by clearings, plazas, and courtyards with sculpture, ornate benches, urns, and water displayed in many ways.

Monumento a la Ciudad de México: A Portal through Time and Place

If the Urrutia arch is a reflection of diverse influences,[24] Monumento a la Ciudad de México, the entrance gate on Hildebrand, honors just one influence on Mexico—Spain. Flanking the wrought-iron, double-width gate are two imposing, mirror-image entry towers. Capped by finials and ornate wrought-iron and colored glass lanterns, the towers together display four nearly identical Talavera tile murals, one each on the interior and exterior walls. Outside the walls, the murals beckon toward the nearby chapel of the Incarnate Word, as water dribbles through twin lion's head fountain spigots into half-moon troughs to welcome thirsty travelers.

Above each pool, a colorful riot of flowers and greenery flows from a blue vase sitting on a stylized golden table, bordered by a vibrant blue and yellow pattern with a white edge. The lion's head repeats at the top, declaring that this entrance commemorates the "IV Centennial" and elaborating that Hernán Cortés founded Mexico City in 1521 and that four centuries later, in 1921, Urrutia founded this institution (Miraflores). Each mural is signed by Marcelo Izaguirre, a Mexican architect and civil engineer, and brother of Baltazar Izaguirre Rojo, a friend of Urrutia's and a well-known Mexican poet. The striking flower vase design may be familiar to one who had once, for example, wandered the streets of Churubusco, a town near Coyoacán. Urrutia knew this place well, and in the Church of Santa Maria de los Angeles (1698) at the turn of the century, a tile panel remarkably similar to Urrutia's (all the way down to its bordered pattern) resided in the entrance to the choir balcony.[25]

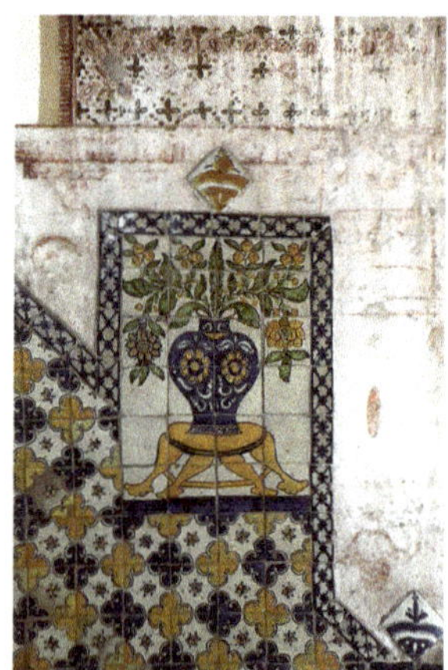
Talavera panel, Ex-Convento de Churubusco.

History connects to the present time through art. The IV Centennial, for instance, bridges four centuries from the conquest of Mexico to Urrutia's creation of the garden. And now, though he is gone, he leaves this portal from which to view a river of complex

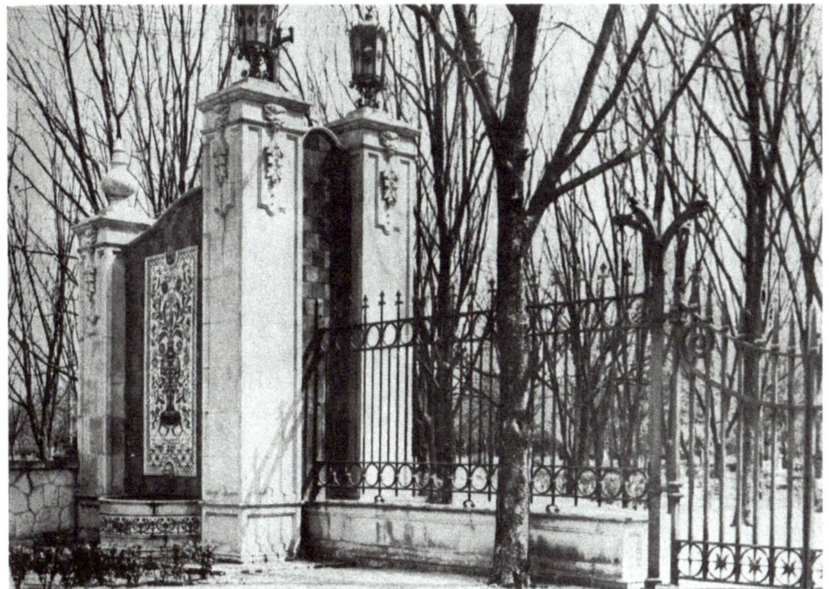

Monumento a la Ciudad de México, east tower.

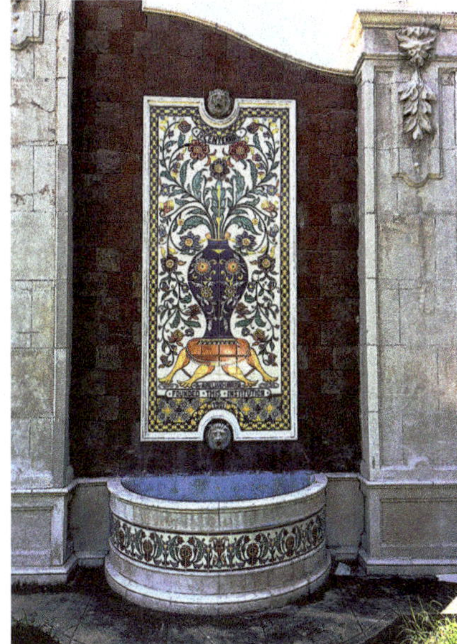

Monumento a la Ciudad de México, detail.

waters flowing from Mexico, rich with history and culture. Hernán Cortés's conquest of Mexico should be neither forgotten nor glorified. Spain's invasion was violent, but the spears and arrows of the Spanish at this gate become the beautiful wrought-iron boundary to a flowering Mexican garden, equally beautiful in San Antonio as it is in Churubusco. Mexico is inseparable from this reality.

Calzada del 2 de Abril: The Díaz Road to Success

»The canopied, tree-lined entry drive ushers us to a taller-than-human square column topped with a sculptural monument to Porfirio Díaz positioned in front of an imposing sentry tower.« As Urrutia saw it, Díaz (1830–1915) almost singly directed Urrutia's modest beginnings into a starring role in advances in the field of twentieth-century surgery.

»Punctuating the drive as it moves toward the bust of Díaz are four other square columns, two on either side of the drive, commemorating the first fifty years of Urrutia's flourishing practice (from 1895 in Mexico City to 1945 in San Antonio), along with two pedestals with large cuplike (*taza*) forms.[26] The columns are a generous gift from his

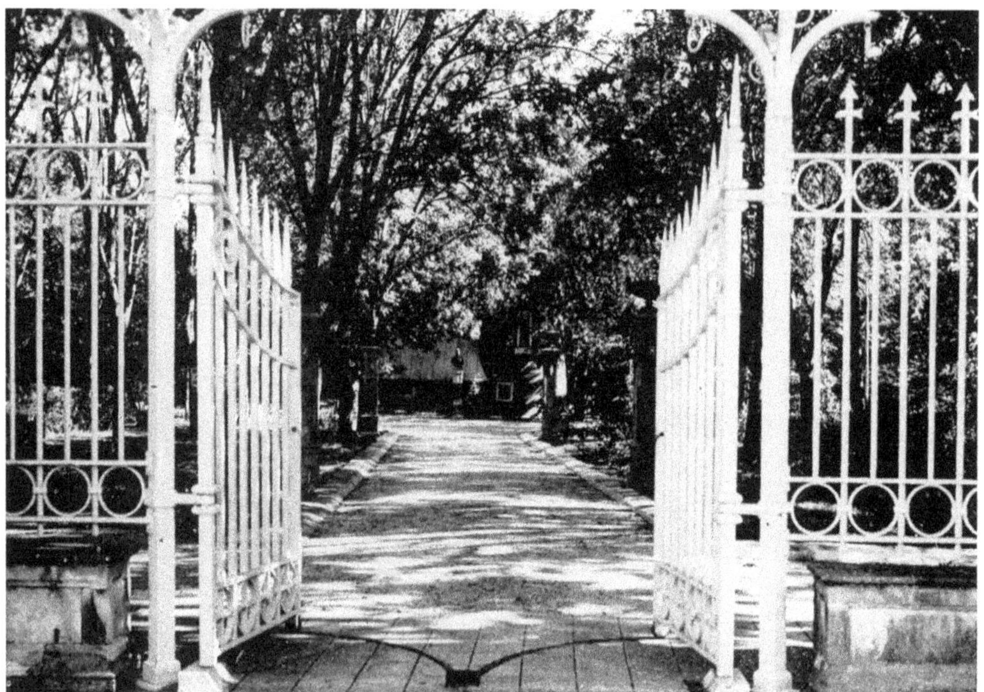

Calzada del 2 de Abril,
12 x 165 feet, nonextant.

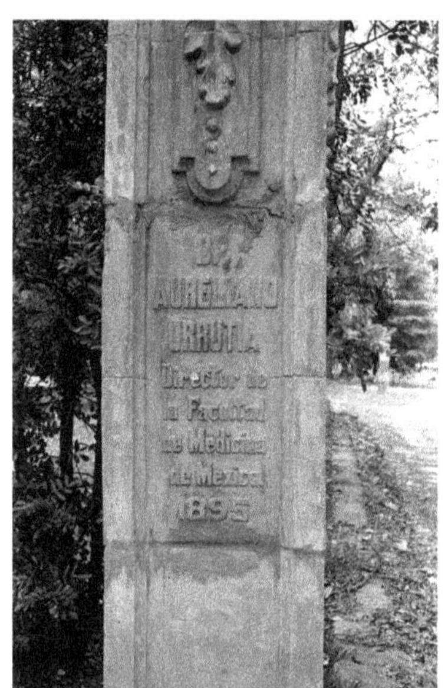

Commemorative column, details,
8 x 2 x 2 feet.

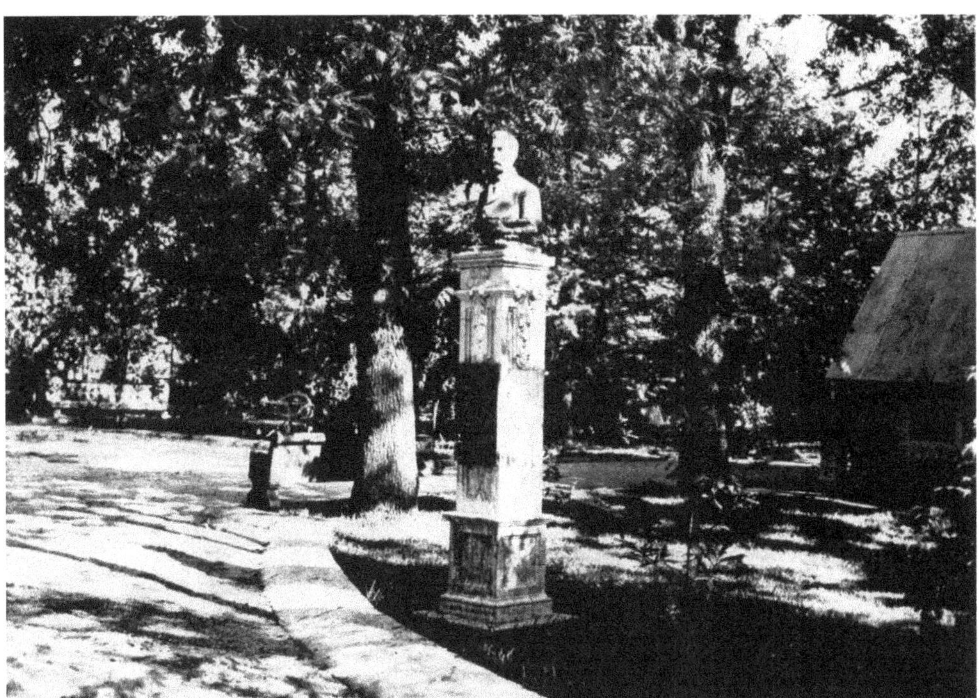
Monumento del General Porfirio Díaz, 3 x 3 x 2 feet exclusive of base, nonextant.

children, placed late in the garden's heyday.« Urrutia would practice for another fifteen years, until his retirement in 1960 at age eighty-eight.

»Urrutia often named the objects in his domain, and the entry drive was no exception. The drive, named Calzada del 2 de Abril in remembrance of Díaz, transports us to Díaz's most famous accomplishment, the final Battle of Puebla, which occurred on April 2, 1867.« The young general emerged victorious, paving the way for the Republic of Mexico to decisively retake control of the country from French rule in the Second Franco-Mexican War.[27]

Díaz was president of Mexico for nearly Urrutia's entire life before exile. A protégé of the previous president, Benito Juárez, the young lawyer-turned-general became president in 1877 when Urrutia was almost five years old and essentially held the office for three decades, until 1911. Continuing the trend of success among the indigenous class characterized by both Juárez and Díaz, Urrutia benefited from Díaz's educational initiatives for those of lower socioeconomic status. After primary school in Xochimilco, Urrutia attended high school at the prestigious Escuela Nacional Preparatoria in Mexico City.[28]

From early education to graduation, Urrutia's scholarly prowess gained him national recognition. Díaz personally granted him an award as top student upon his graduation from high school, opening the door to Urrutia's medical career.

Two men linked together by accomplishment, each man reflected the most humane aspirations of the other. Urrutia relished Díaz's later visit to the Sanatorio Urrutia in Coyoacán, recalling how Díaz proudly retold the story of awarding Urrutia with the scholar's prize.[29] In the early twentieth century, Urrutia's pioneering surgical work prompted the president to commission photographer Guillermo Kahlo[30] to film Urrutia in the surgical theater in order to showcase Mexico's progress in the medical industry.[31] »At Miraflores, Urrutia further memorialized Díaz in a plaque below the bust that reads "*el más Grande de los Grandes hombres. Hizo la Patria*" (the greatest of great men, who made the fatherland).«

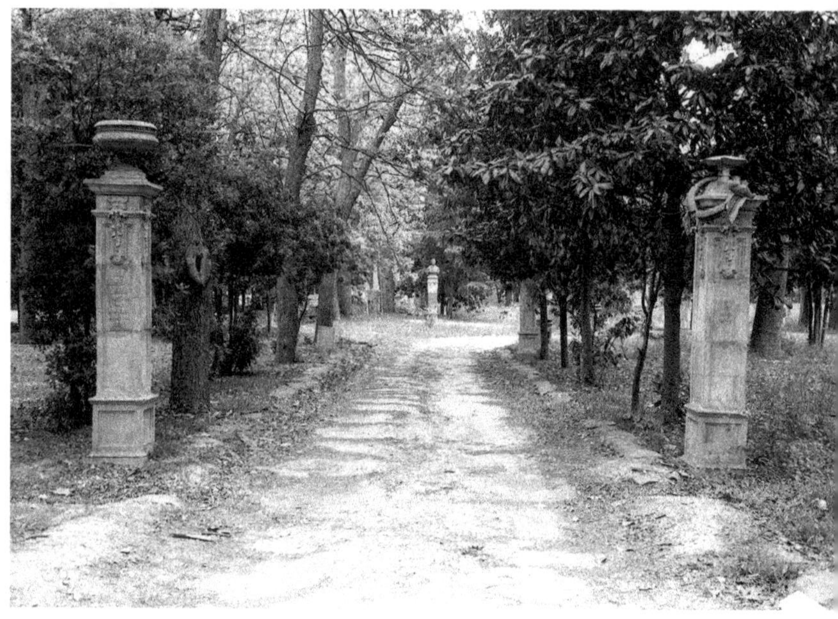

Calzada with Monumento del Díaz bust.

Urrutia epitomizes the ideals of industrial and economic progress initiated by Díaz. But in the long run, Díaz's failure to positively affect enough people led toward the Mexican Revolution. Was it lost on Urrutia that the man he idolized, unable to secure viable presidential succession, unwittingly contributed to his exile? »Urrutia pays homage to Díaz, and the existence of the garden indicates both the successes and failures of this important Mexican leader. After all, it is through exile that Miraflores became the only Mexican garden outside of Mexico.«

For the Love of Talavera

»Inside the Monumento a la Ciudad de México gate, the forested landscape reveals a panorama of color in the near distance. Straight ahead, behind the Díaz bust, a single-story reddish stone pavilion abuts a distinctive three-story circular tower.« To the west stands a majestic blue and white bench. »Below it lies a shallow quatrefoil pool.« To the southeast, a massive orange, green, and yellow triple-tiered bench glimmers through the trees. »Nearby, water erupts from a towering fountain resembling a volcanic rock formation.«

The two benches, perceived from afar as sculptural objects, are, upon closer inspection, clothed in hundreds of colorful traditional Mexican tiles, or *azulejos*. »Throughout the garden, more of these colorful benches await, each one unique, presenting its own intimate view of the garden, but all inviting a visitor to sit and enjoy.«

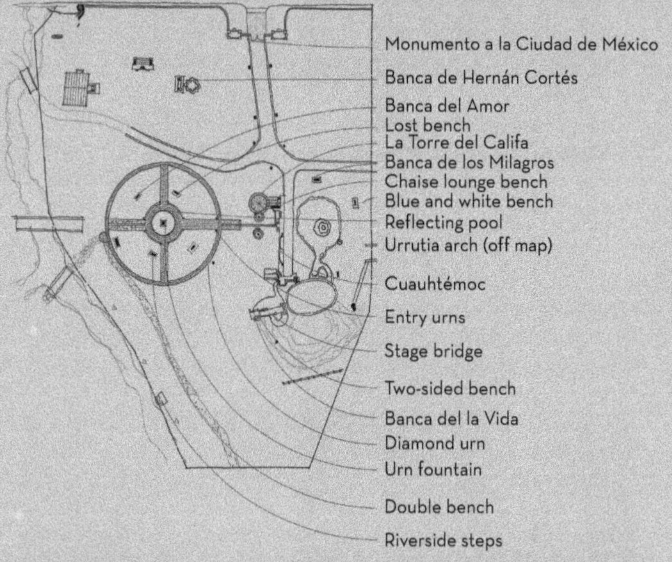

Clay and Color: A Bridge between Men

In 1920s San Antonio, only one person loved colorful Mexican tile as much as Urrutia—Atlee B. Ayres. After his graduation from architecture school in Manhattan, Ayres took his first job in Mexico City in 1896, working a year for architect George King. Perhaps he had occasion to sit on the prerevolutionary Talavera benches at Urrutia's Xochimilco hacienda, but more likely, the two men met stateside.

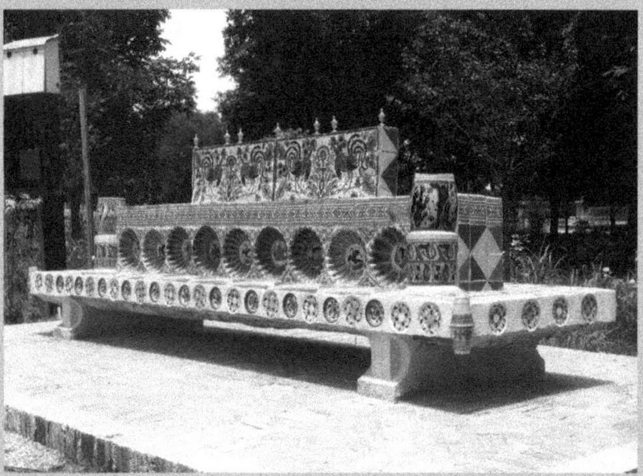

Banca de los Milagros, 5 x 15 x 4 feet, now in ruins.

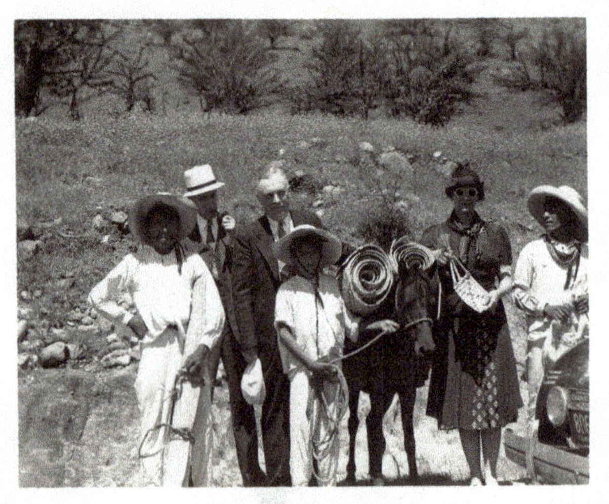

Atlee B. Ayres, center, with George Willis, back left, and Ayres's wife Olive Moss, far right, with indigenous craftsmen in Mexico, 1925.

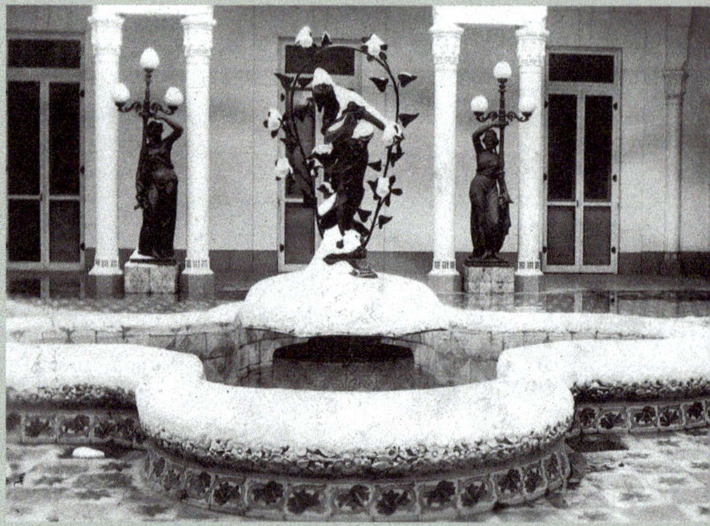

Interior courtyard, Quinta Urrutia, nonextant.

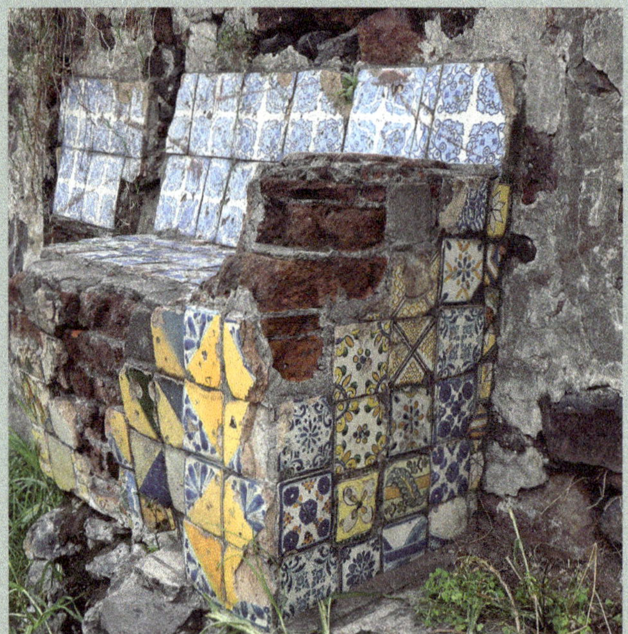

Mexican bench, brick and Talavera, 3 x 5 x 2 feet, built ca. 1900, Xochimilco, Mexico, shown in 2018.

In 1918, Ayres certainly saw Urrutia's stunning Spanish-style mansion, Quinta Urrutia, with its ten-thousand-square-foot footprint, going up on Broadway. It dwarfed the nearby conventional wood-frame bungalows of Mahncke Park and made a statement of Mexican residential style unprecedented in San Antonio at the time. Brackenridge Park served as a backdrop to this relatively undeveloped stretch of River Avenue. Mexican tile was used indoors and out at the house, from its beautifully patterned pressed concrete tile floors to its Talavera wall murals and decorative mosaic features.

So great was Ayres's love for Mexico, and so resourceful and social was he, that it would have been out of character for him not to meet Urrutia or the Quinta Urrutia architect, Porfirio Treviño, of Monterrey, Mexico. Ayres, who was just ending his stint as state architect, was getting ready to enter the prime of his architectural career in San Antonio. He would have

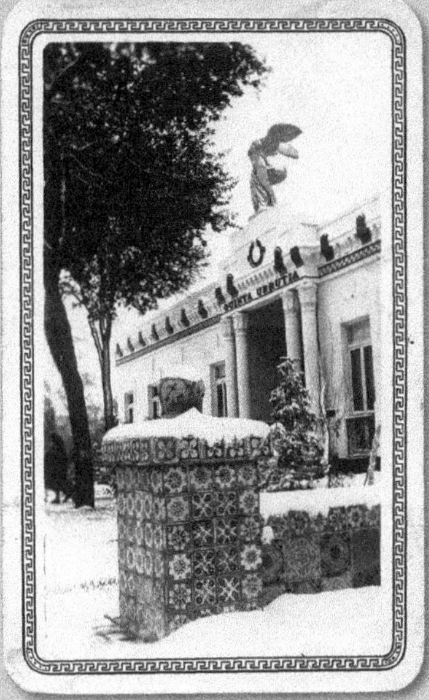

Low seatwall, Quinta Urrutia.

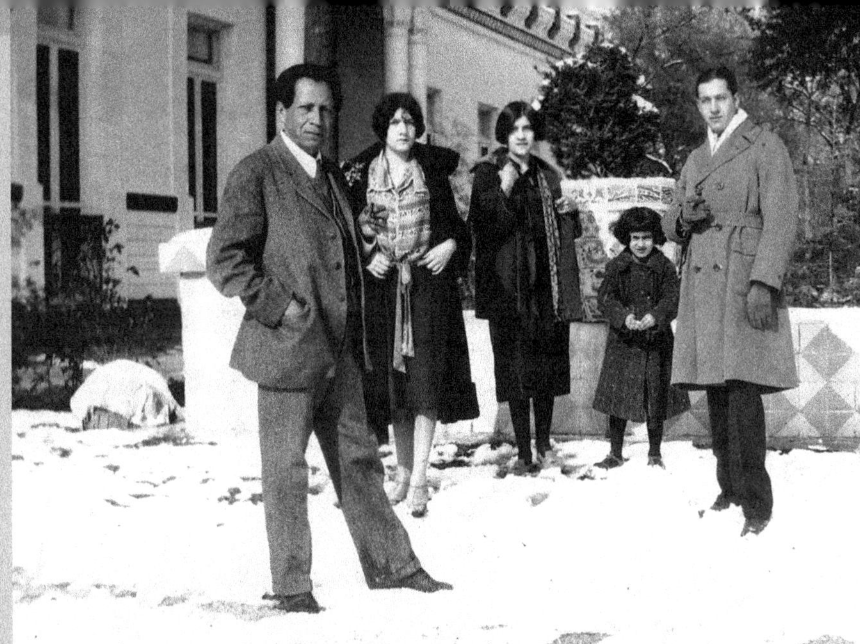

Urrutia, left, with family members, Quinta Urrutia, ca. 1926.

noticed Urrutia's collection of Talavera seatwalls, placed at intervals in the Quinta Urrutia garden, near the entrance to the house. These rectangular structures were freestanding in the lawn, capped at the ends by slightly taller columns or punctuated by an urn planter perched at the center. Formed from brick and concrete, and clad in hundreds of colorful *azulejos*, the seats, more than two feet high and fifteen feet long, accommodated eight, ten, or twelve people and often served as a background for photographs of the Urrutia family as they came and went from the home.

Urrutia purchased the land for Miraflores in 1921 and began to make Talavera tilework a unifying feature of the garden by using it extensively in the two entrance gates at Hildebrand and Broadway. Urrutia relocated at least three massive seatwalls from Quinta Urrutia to Miraflores, a mile away, repurposing them into larger and grander raised bench designs. The wall of each of the house's seats became the backrest to a new bench at Miraflores, where it was lifted onto a wider horizontal slab seat set on legs.

Urrutia went on to design at least eight unique Talavera benches to carry the Talavera theme throughout the garden. Some were freestanding, others were incorporated into the architecture, and still others were nestled into more naturalistic locations. The benches added a ribbon of color and unity to the garden, and gave visitors ample places to sit and contemplate the natural beauty around them.

Ayres became quite obsessed with Urrutia's collection of Talavera tiles, some of them antiques, and Mexican tile in general. Urrutia's use of many types of Mexican tile in his home on Broadway and in the gates and benches of Miraflores was highly unusual, even unprecedented, in San Antonio, which was not employing colorful Mexican tile or white stucco in its homes at the time. In the early

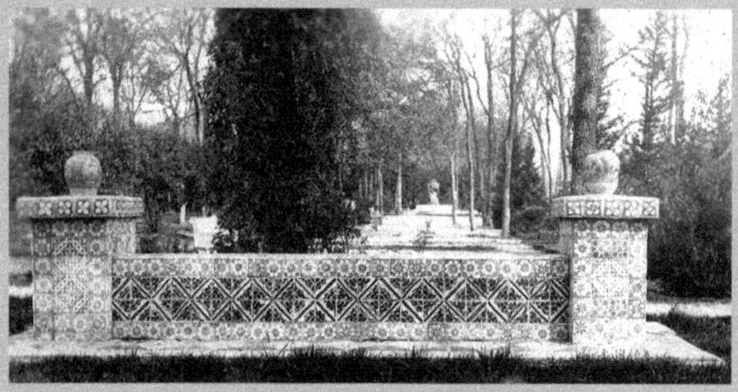

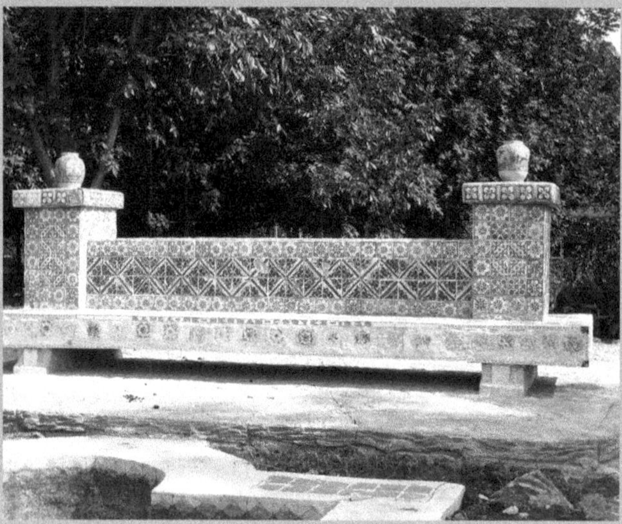

At top, seatwall, Quinta Urrutia; below, reconfigured as a bench at Miraflores.

to mid-1920s, the Talavera workshops of Puebla were emerging from a long period of decline, and their work was rare, expensive, and collectible. The origins and history of Talavera tile in colonial Mexico had only recently been academically explored, as the Philadelphia Museum and the Metropolitan Museum of Art in New York established collections of it between 1906 and 1916.[32]

The rebirth of the Talavera industry allowed Urrutia and Ayres to delve full force into designing with tile, each in their own way. As Urrutia built his home in 1918, Ayres traveled to California to learn about the blossoming Spanish Colonial Revival homebuilding trend. And as Urrutia incorporated his Talavera collection into Miraflores between 1921 and 1925, Ayres traveled to Spain and then interior Mexico to work on his 1926 book *Mexican Architecture*.[33] During his time in Mexico Ayres procured a photograph of the new tile benches of the Fuente (fountain) del Quijote in the Bosque de Chapultepec, a large park in Mexico City, in addition to documenting many other uses of tile for his book.[34]

Today, if you go to the McNay Art Museum, in the beautiful interior courtyard you will see a tile peacock wall panel from the Uriarte Talavera workshop in Puebla, used by Ayres in 1927. It is similar to the one Urrutia used on the Urrutia arch at Miraflores in 1924. Ayres also incorporated tile benches of his own design into a few of his Spanish Colonial Revival homes.

In 1930, as business slowed during the Great Depression, Ayres became more focused on his photography hobby. One Sunday he set out to photograph the tile seats at Miraflores, in order to try to sell his images to a publisher.[35] In a pitch introducing himself as "an amateur photographer" who had recently gone "out on a little photographing tour," he wrote: "We have in San Antonio a man by the name of Dr. Urrutia, who came here from the City of Mexico some few years ago. He has quite a good size piece of property on the outskirts of town and has built up a lot of pools, gardens, etc. . . . He is quite an artistic fellow and collected a lot of beautiful antique tile which he has used in a most interesting way in building up seats around his pools, fountains, etc. They are most colorful and original."[36]

Urrutia collaborated with Ayres as the architect tried to place the photographs with five architecture magazines. The doctor created detailed descriptions and even fable-like stories about the various seats.[37] Their work, however, was rejected.

Then, in March 1931, Ayres was appointed to the committee tasked with hosting the annual American Institute of Architects (AIA) convention to be held that April in San Antonio. Mrs. Ayres (née Olive Moss) approached Urrutia to ask if the spouses of the architects could visit Miraflores. He not only enthusiastically agreed but also invited the entire convention for an evening fiesta at Miraflores. The convention leadership also toured Quinta Urrutia. The architects pronounced the party—which featured performances by costumed Mexican dancers, singers, and a string orchestra, followed by a dinner of traditional Mexican food—"one of the most interesting and colorful that they had ever attended." Indeed, most of the attendees, who came from all over the United States, had never before been exposed to any aspect of Mexican culture. The tradition of Mexican fiesta was relatively unknown in the United States at that time. A correspondent from *American Architect* magazine commented to Ayres that he would "remember the Convention . . . long after many other conventions [were] forgotten."

The institute presented Urrutia with a hand-illuminated parchment scroll in thanks. That year,

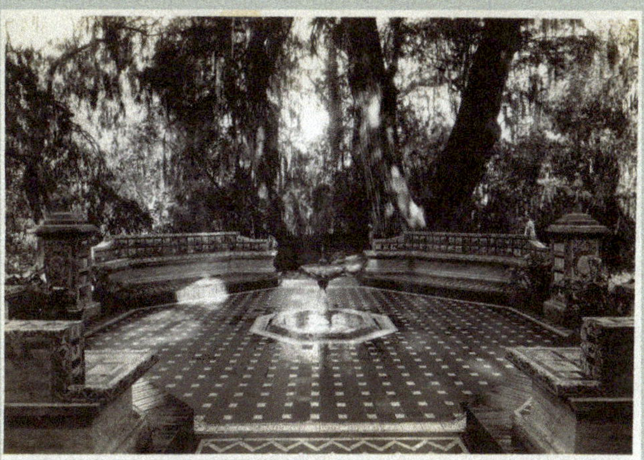

Fuente del Quijote, Bosque de Chapultepec, Mexico City, ca. 1925.

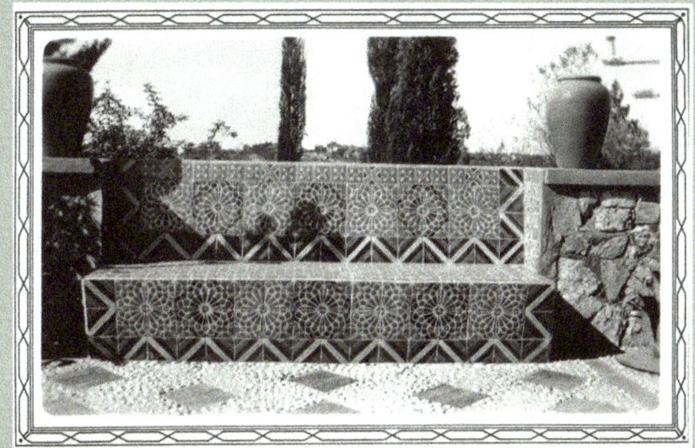

McNay Art Museum tile bench, Atlee B. Ayres, ca. 1929, nonextant.

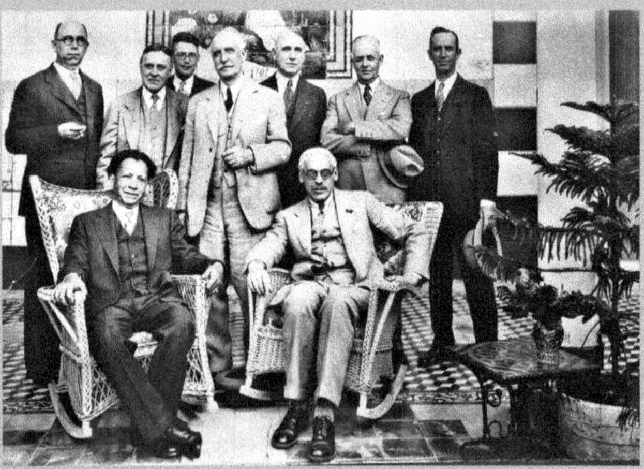

American Institute of Architects leadership visit with Urrutia, bottom left. Atlee B. Ayres stands second from right. Quinta Urrutia, 1931.

riding on the success and recognition of the convention, Ayres became the first San Antonio architect to receive an AIA fellowship.[38] Shortly thereafter Ayres received an invitation to publish his photographs of Urrutia's benches in *California Arts and Architecture*. Because of the Depression, the magazine could only offer Ayres a two-year subscription in compensation. After more than two years of pitching the photos, Urrutia saw his eight tile benches, photographed by his co-enthusiast Atlee B. Ayres, published in 1932 in a striking two-page spread.[39]

Although each man was impressive in his respective field, together the two went above and beyond to build an early bridge between their cultures. Urrutia's bold expressions of Quinta Urrutia and Miraflores at least partially inspired Ayres to move toward an architectural style that changed San Antonio. Ayres's act of publishing Urrutia's benches and showcasing his home and garden to architects nationwide acknowledged Urrutia's influence on his thinking. Perhaps even more importantly, their shared love of Mexico and their creativity allowed them to find a way to collaborate. Each validated the other's thoughts that San Antonio's connection with Mexico was important, and that if they built on it they could reconnect to Mexico in a special way to create places of unique beauty.

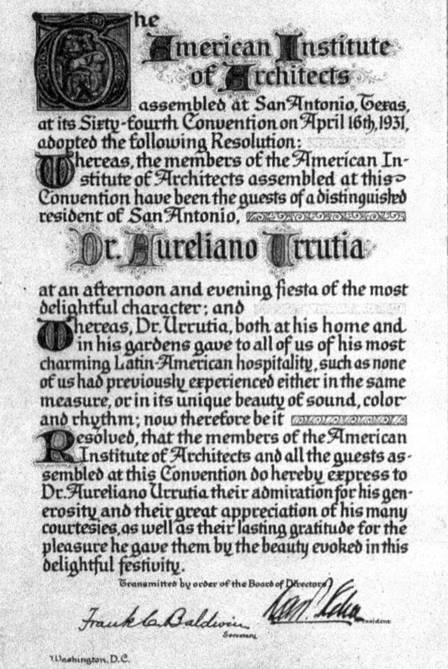

The American Institute of Architects assembled at San Antonio, Texas, at its Sixty-fourth Convention on April 16th, 1931, adopted the following Resolution:

Whereas, the members of the American Institute of Architects assembled at this Convention have been the guests of a distinguished resident of San Antonio,

Dr. Aureliano Urrutia

at an afternoon and evening fiesta of the most delightful character; and

Whereas, Dr. Urrutia, both at his home and in his gardens gave to all of us of his most charming Latin-American hospitality, such as none of us had previously experienced either in the same measure, or in its unique beauty of sound, color and rhythm; now therefore be it

Resolved, that the members of the American Institute of Architects and all the guests assembled at this Convention do hereby express to Dr. Aureliano Urrutia their admiration for his generosity and their great appreciation of his many courtesies, as well as their lasting gratitude for the pleasure he gave them by the beauty evoked in this delightful festivity.

Transmitted by order of the Board of Directors

Frank C. Baldwin, Secretary

President

Washington, D.C.

Hand-illuminated scroll, 1931.

18 *California* ARTS & ARCHITECTURE

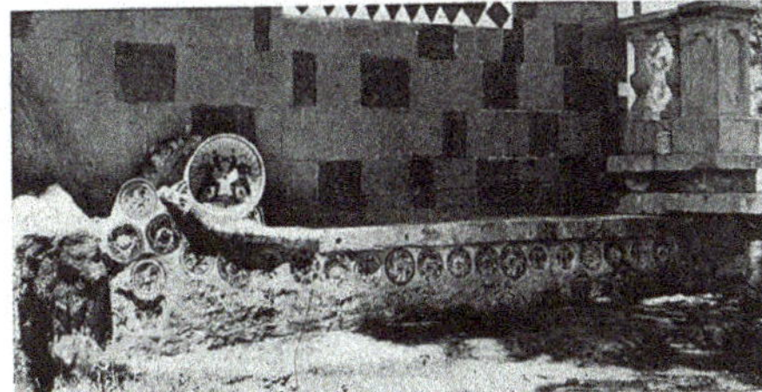

On the wall of a pavilion at "Mira Flores", Dr. Urritia's estate, is a reclining seat in which Mexican tile plates were used. Below, an amazing bench at whose ends, like columns, are barrels used for water by great families during the 17th century; above is Espano Araby tile from Talavera de la Reina with genuine Persian design, in blue and white; the deep panels are gold, red and green. Lower left, a "Banco del Amor" with two sides, one for the weeping Princess, the other for her singing lover; its tiles are blue and white with touches of light yellow. Lower right, "La Banca del Esterio Chino", cooking dishes brought from the Philippines by the Spaniards, as was also the weeping willow tree.

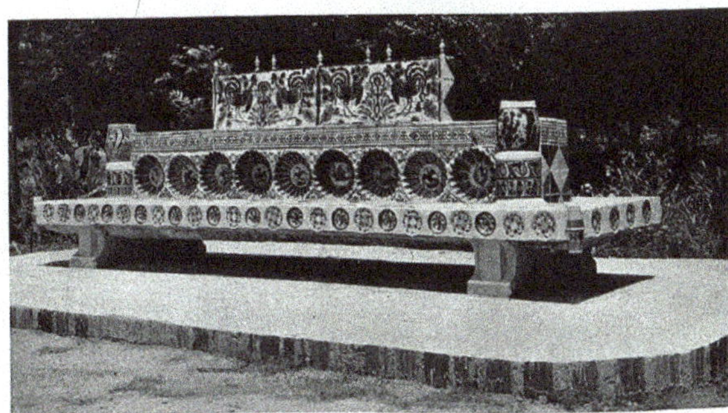

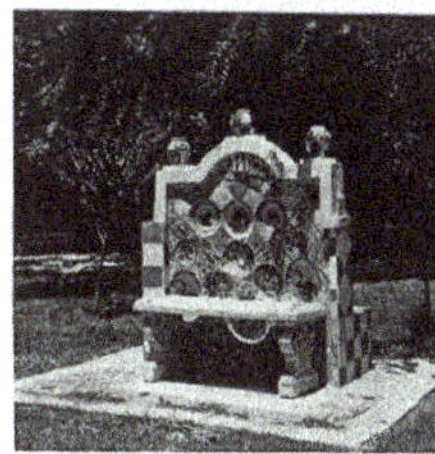

There is a legend that a Chinese Princess married a Spanish Governor of the Philippines, and they migrated to Mexico, bringing a potter who taught the Mexicans the color and design of Chinese tile—whose influence is clearly seen in Mexican tile.

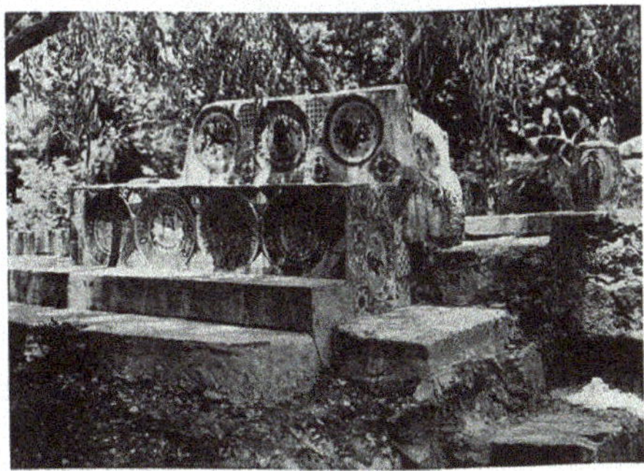

June, 1932 19

For the country place of Dr. Aurelino Urritia, Atlee B. Ayres, architect, has utilized a remarkable collection of old tile in various garden seats. Right, a bench built by Don Cortez in 1535, of tile brought from Spain for his Cuernavaca palace. Below, the rarest collection of tile ever produced; used in the house of Olif the Conqueror, of tile taken from Moorish castles in Guadalupe. Lower left, the "Banca de la Vida" (a place to forget everything) with Cortez' Coat of Arms and tile from Granada. Lower right, a double bench for the Califa and his wife, tile taken from a fountain in the Convent de la Merced.

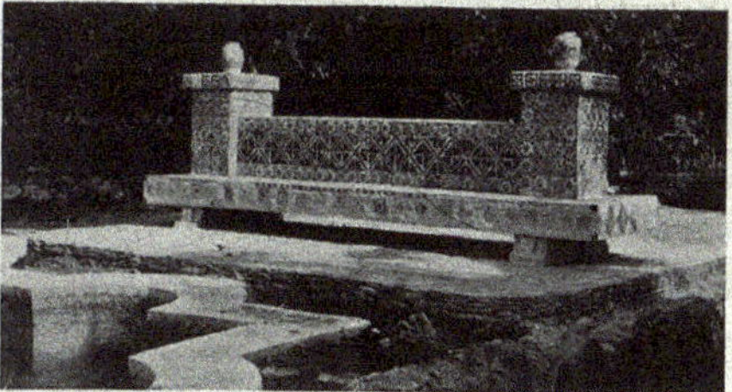

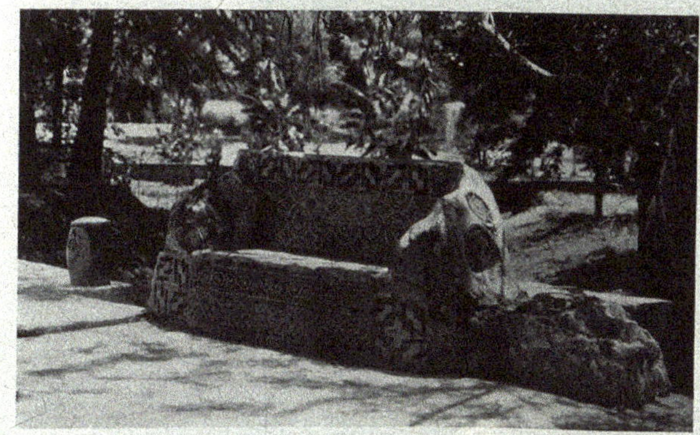

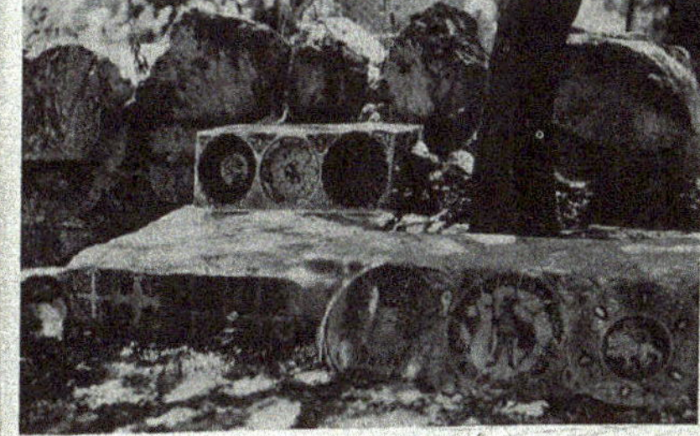

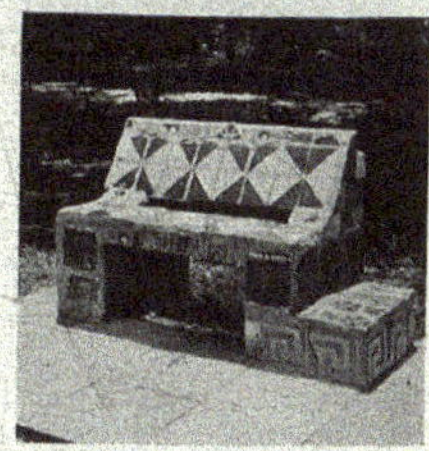

These tiles are mainly in blue and white, but with black, yellow and green in panels or plates.

THE USE OF OLD SPANISH AND MEXICAN TILES ON DR. URRETIA'S ESTATE, SAN ANTONIO, TEXAS

California Arts and Architecture, center spread, October 1932.

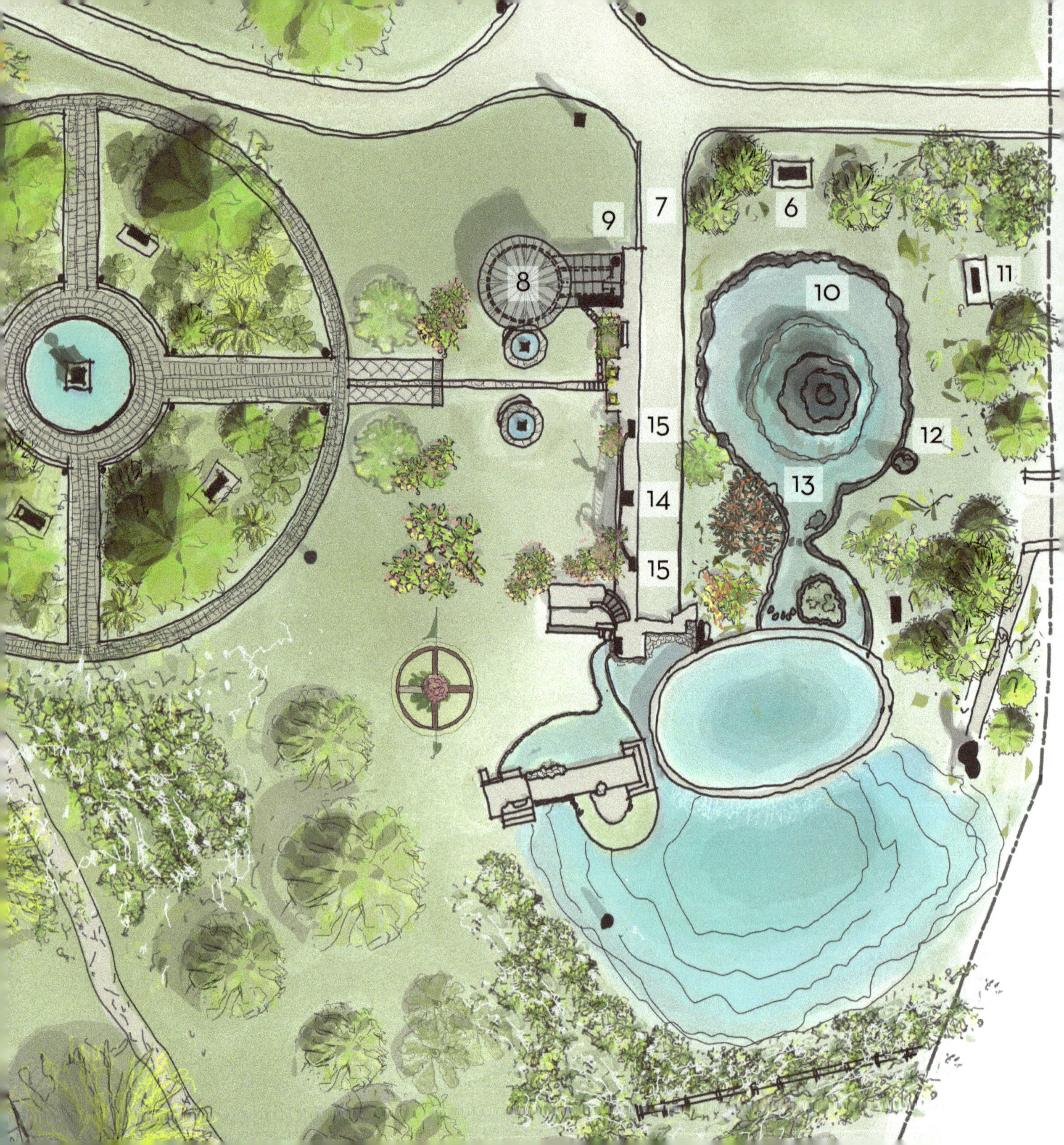
9
7
6
8
11
10
15
12
13
14
15

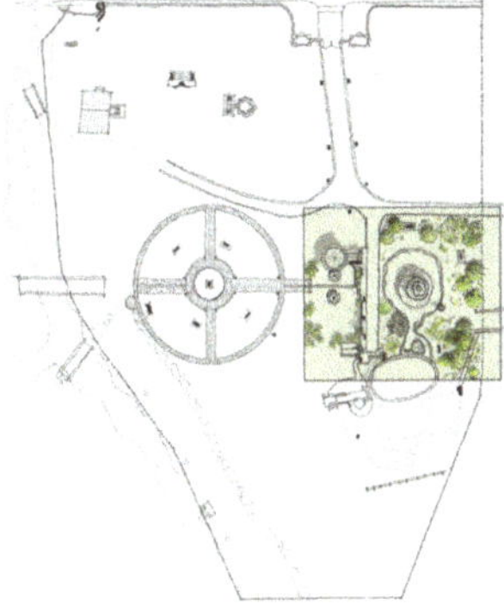

The Esplanade of Time

»The tower pavilion behind the bust of Díaz beckons one to a one-hundred-foot esplanade revealing the garden's heart.« At the gates to Miraflores, Urrutia introduced symbols referencing Spanish conquest, Catholic education, and Mexican independence, with a hint of allusion to the Aztec past with depictions of jaguar and cypress. But deeper in the garden, familiar history recedes as the esplanade ultimately flows toward references to the origin stories of preconquest Mesoamericans, a people who exist in collective Mexican memory, passed along through ancient language and myth. »Facing south, with the tower entrance at right, a large fountain erupts from the earth at left, its water gushing and then flowing south alongside the walk.«

6 Banca de los Milagros
7 Esplanade
8 La Torre del Califa
9 Hollow tree stump utility column
10 La Fuente Monumental
11 Blue and white bench
12 Ana María grotto
13 Rivulets
14 Cuauhtémoc
15 Eagle columns

Esplanade overview, shown ca. 1952.

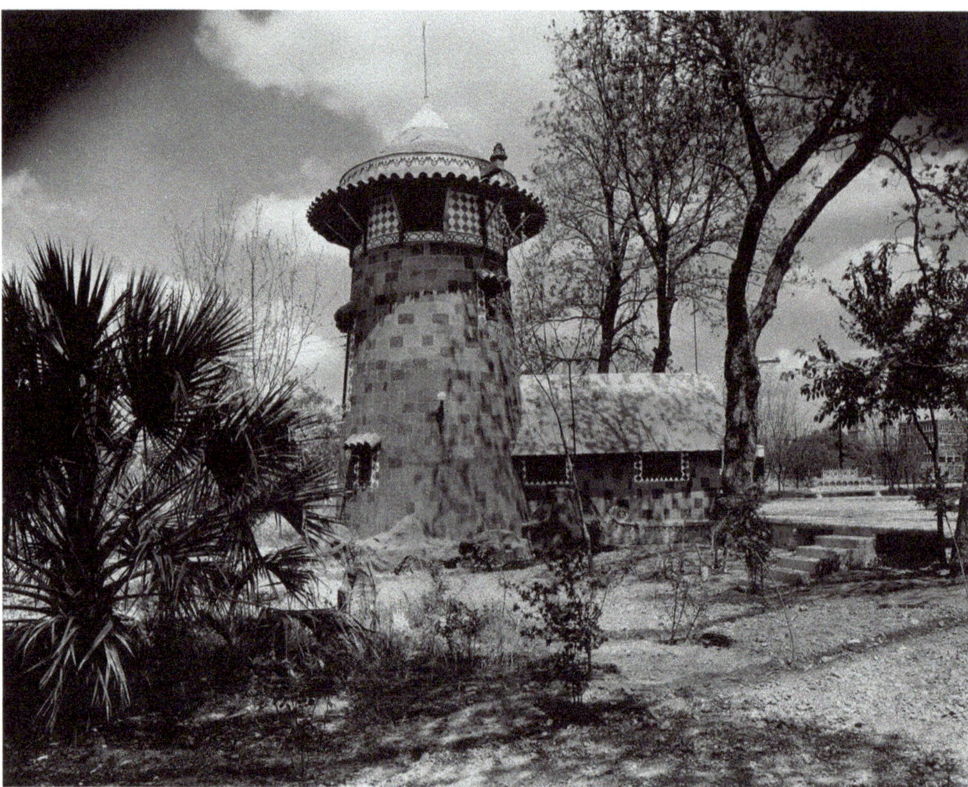

La Torre del Califa, 20 feet in diameter, with a one-story entry, nonextant.

La Torre del Califa: An Elevated View

»A watchtower, which Urrutia or perhaps his poet friend Tablada named La Torre del Califa, is capped with a cross at the peak of its conical roof and another above its door; the entrance is situated at the end of a long, narrow, single-story room, which serves as the tower's entry hall. The windows in the three-story tower promise new views of the path already traveled and of future destinations.«

»Massive faux stone boulders anchor the fortress observatory.[40] Above them the tower, its walls clad in a reddish stone veneer, narrows slightly as it reaches a single circular room at the top, sheathed on the outside by diamond-patterned tiles checkered blue and white. Layered, colorful metal bands form scalloped edges on the playful, deeply overhanging roof. Angled wooden beams buttress the eaves, with a horizontal beam jutting out to hold a sounding bell roped to a window ledge, ready to ring out over the property.«

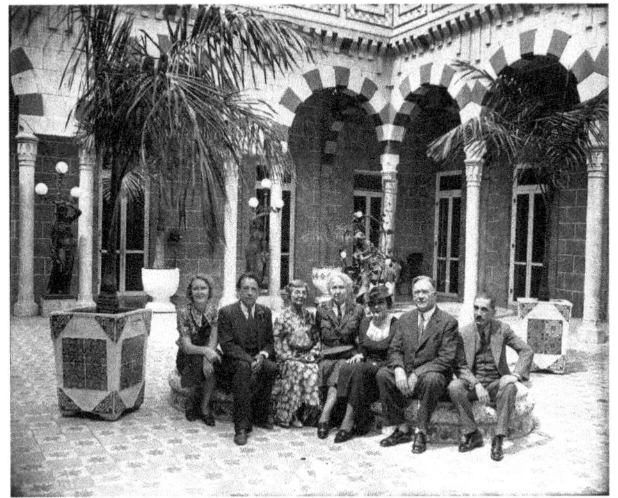

Quinta Urrutia, interior courtyard, 1934, nonextant.

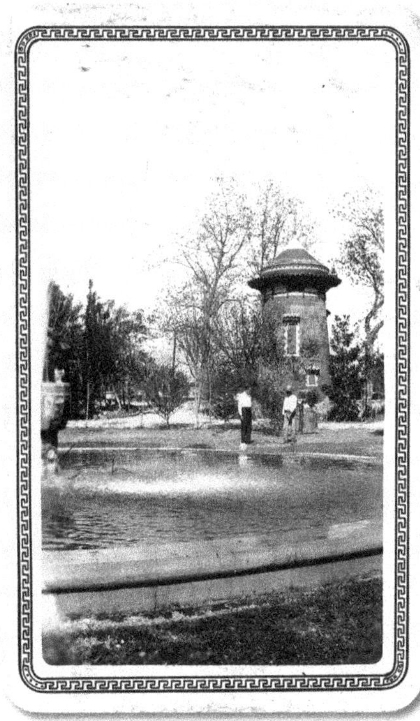

View across reflecting pool toward the Torre del Califa, with Urrutia, center left.

At ground level outside the tower, a hollow faux bois tree stump serves as an ingenious housing for electrical equipment.[41] »Through the entrance next to the stump and past a long narrow foyer, a spiral staircase ascends, running along the circular wall. Windows punctuate the view in all directions at rising heights. A windowed library room at the top holds a collection of books—a sanctuary of sorts for reading, relaxing, and contemplation, even while the tower affords a commanding view under the protection of the jaguars and reclaimed Spanish spears on the perimeter fence.«

La Torre del Califa, as hinted at by the Talavera entry gates, evokes the beauty, intricacy, and serenity of Islamic influence in a building style brought by Spain to Mexico. Urrutia experienced such architecture and design daily in Mexico, and appreciated the influence of the Moors and Islamic art on Spanish architecture.[42] Like Miraflores, the Sanatorio Urrutia in Coyoacán was also guarded by a sentry tower, which looked very much like a Moorish tower. Urrutia had also incorporated elements of Islamic (Mozarabic) style into his home, Quinta Urrutia, a mile south on Broadway. Yet beyond the influence of Cordova, Seville, or even Mexico City or Xochimilco, these structures and others in the garden carry a uniqueness all their own: a fanciful yet sophisticated style that illuminates the multifaceted and complex nature of Urrutia.

Tablada writes in a letter to Urrutia that the tower reminds him of the Islamic proverb "Make a child, plant a tree, write a book." As a father, Tablada says, Urrutia has given "his numerous offspring not only physical life, but has set an example of admirable behavior." The trees he planted "give shade and health to bodies and serene joy to souls." And the surgeon's scalpel is "without equal," adding "pages" to the book of life by excising that which would cause "pain and death!"[43]

For Tablada, the tower represents the Islamic principle of an ethical existence lived through having children, nurturing, and actively working to better people's lives. He sees Urrutia fulfilling these aspirations and Miraflores as a reflection of these values. Practically speaking, towers serve multiple purposes—a call to prayer, a protective vantage point, a solitary roost, an elevated view. And, for Urrutia, this intriguing space is a place of solitude and study that balances looking in and looking out.

Banca de los Milagros: A Grand Place to Dwell

»A variety of objects, tied to Mexico's colonial heritage, are clustered around a tiered fountain, which evokes the volcanic landscape surrounding Mexico City.« Most prominently, as if perched on an Aztec lakeside, an ornate Talavera bench sits like a gigantic, fanciful tiered wedding cake. The massive seat easily hosts a crowd of two dozen or more. Its variety of colors and tiles, not all of them Talavera, gives the impression of a vast and experimental mixing. It is an apt resting place from which to depart to places unknown. Urrutia calls it Banca de los Milagros, a "*reliquia de azulejos*" (relic of tiles). He writes: "Made with [the idea of] a Talavera sample book, preserved as a treasure of the seventeenth century. From the Olavarrieta collection, from Puebla. In this sampling of platters and porcelains we discover the Mozarabic influence on the Spanish industry and the Chinese influence on the Poblana industry."[44]

The side of the bench facing the fountain features nine large Talavera serving bowls set vertically into the backrest. Feathered patterns of orange, green, and yellow line the scalloped edges of each bowl, with scenes of nature and civilization decorating each basin center. A blue and white chain border trickles across the top of the arrangement. Large double-stacked water barrels cap the ends, and smaller plates dot the edges of the seat, the corners marked by doubled kissing cups. More large blue and white tiles form diamond patterns on the seat itself. Atop this veritable *cocina* (kitchen) of Talavera a third layer crowns the entire structure—a double panel of crowing roosters, fourteen tiles wide.[45]

Here, the everyday is elevated—the decorative serveware made even more decorative. The tradition of mixing is reflected, layer upon layer—the adoption of the Spanish Talavera industry by the indigenous Mexican population; the combining of the Talavera with other intriguing tiles and constructions in the bench; the exaltation of kitchen

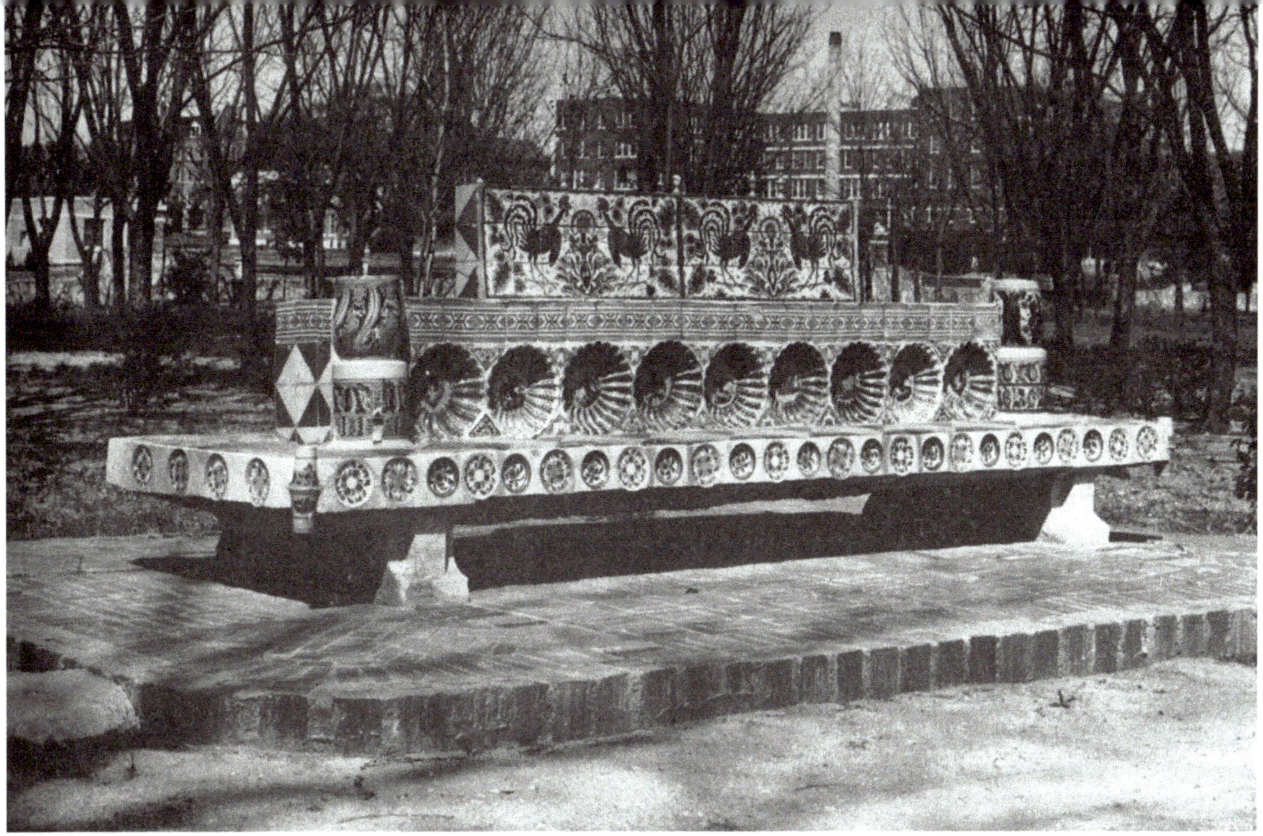

Banca de los Milagros, 5 x 15 x 4 feet, now in ruins.

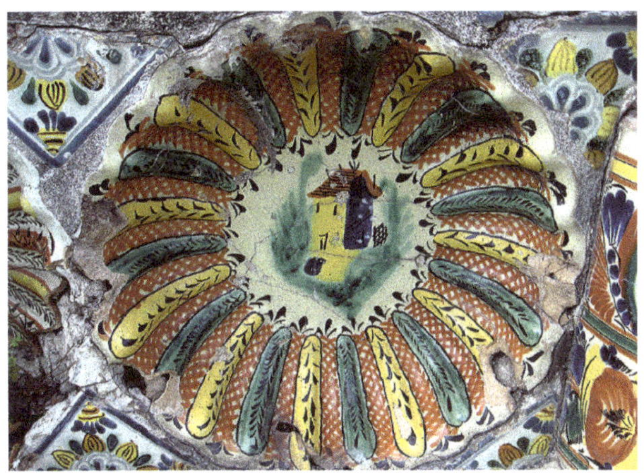
Banca de los Milagros, detail.

objects, accoutrements of a woman's household, crowned by the crowing rooster. Full, proud, and bounteous, this exuberant work of artisanship must also celebrate and remember Urrutia's wife, Luz, the Spanish beauty and stalwart companion and supporter of the exiled man. The bench sits alongside the earthy pyramidal fountain, just as the memory of Luz propels Urrutia, the indigenous man, forward in his life with pride and gratitude.

Ana María Grotto: A Family Remembered

»On the east side of the imposing fountain, a more modest blue and white bench of similar construct to the Banca de los Milagros, decorated with triangular tiles, sits in direct view of the tower. The bench offers a quiet place from which to view the entire scene emerging along the esplanade.« On the southeast edge of the fountain's shallow pool, the Ana María grotto of faux stalactites and stalagmites encloses two putti, one holding a shell-shaped bowl—two children fixed in time.[46]

In 1923, Urrutia married again, to Catalina Tazzer, the daughter of a patient and family friend.[47] Urrutia had originally placed this tribute to the 1925 birth of their second child at the center of a small pool in the garden at Quinta Urrutia.[48] It was a short-lived marriage, and Catalina separated from Urrutia in 1928. Poignantly, Urrutia's first family of eleven children, with Luz, had escaped Mexico in 1914, and now Catalina returned there, pregnant with her and Urrutia's fourth child, planting this branch of the family firmly back on Mexican soil and beyond exile's reach. Moved from its original location of Quinta Urrutia to Miraflores, the grotto commemorates Urrutia's absent Mexican family.[49]

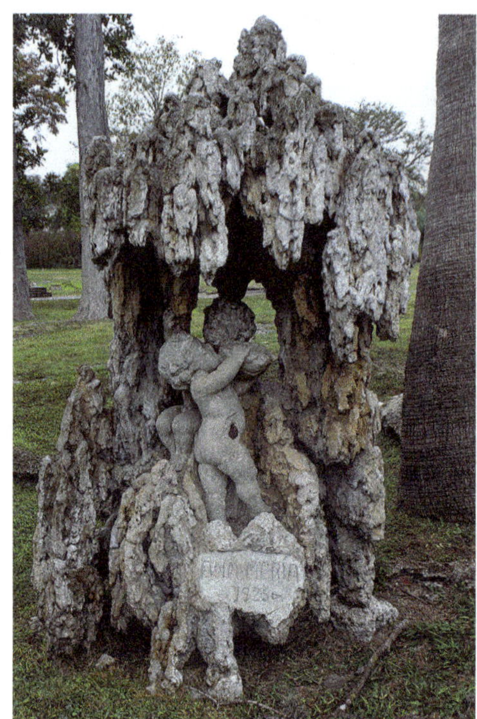

Ana María grotto, *trabajo rústico*, 7 x 4 x 3 feet, restored.

La Fuente Monumental: A Journey Back in Time

As a son of Xochimilco, Urrutia had an affinity for water, which drew him near to the San Antonio River. The river is fed by underground springs, some of which flow beneath the surface of the esplanade, carrying a force only hinted at by the trough fountains at the Hildebrand entrance. »An artesian well forces the water through the gigantic formation of faux volcanic stone rising out of the earth.[50] The stream erupts from the crater, releasing a thundercloud of water that gushes down into a bubbling basin about fifty feet wide and only inches deep, with jagged faux stone edges.[51] Trickling and gurgling southward some eighty feet through a network of rivulets, the water moves onward, corralled by faux stone banks.[52] Small benches offer more resting and viewing places as the water departs.

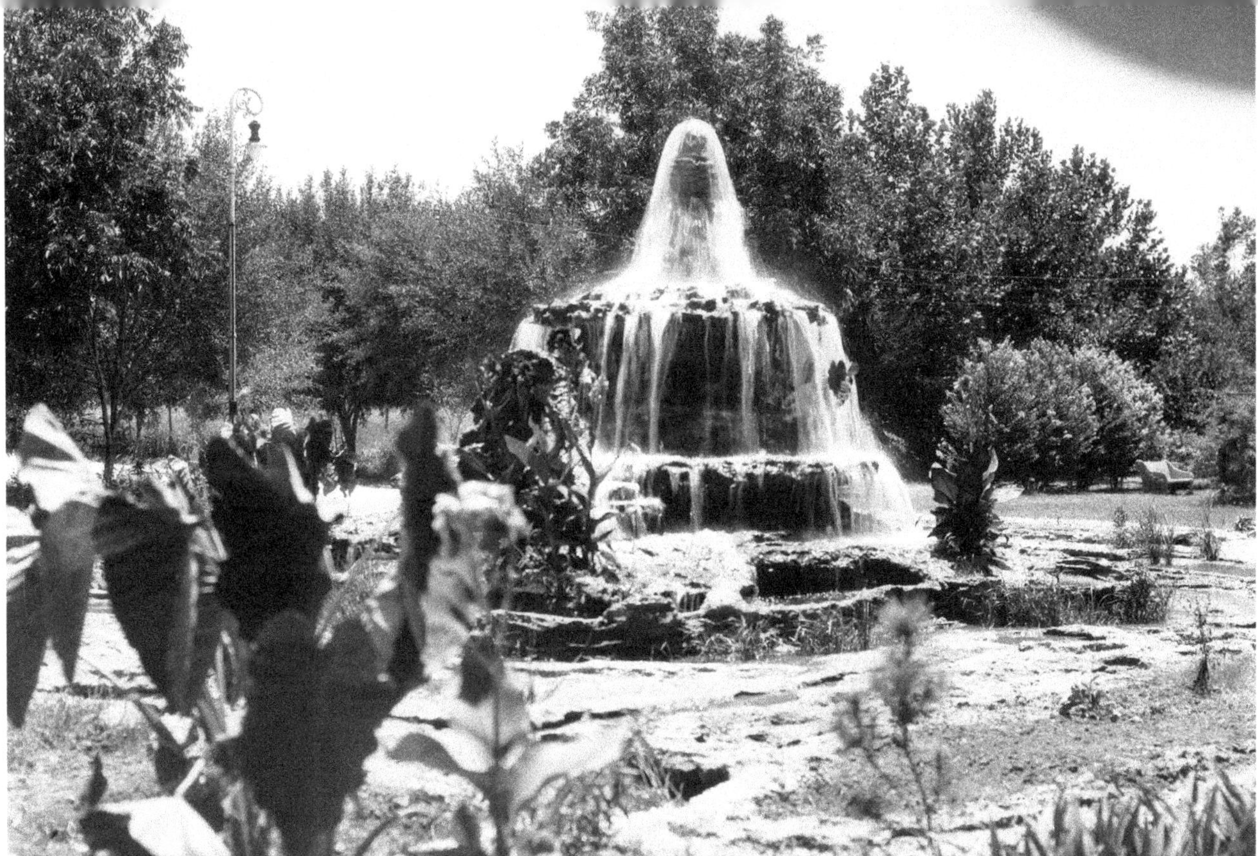

La Fuente Monumental, *trabajo rústico*, 22 x 45 feet, including pool, nonextant.

A chorus of myriad bird songs emanate, as winds rustle the tall trees delivering fragrances of rose, cypress, and magnolia. The shallow channel, lined by Mexican cacti, pours into a large, deep, egg-shaped swimming pool at the southern end of the esplanade.«

Urrutia recalls his birthplace, in both words and this fountain, evoking its nearby volcanoes as the source of life, returning to an ancient time past, the birth of the land itself:

> Xochimilco . . . The most beautiful place in the world; the civilization that ended its beauty had not arrived, and men were not in a position to appraise it. The lagoon of Xochimilco received all the crystalline and pure waters of the thawing majestic volcanoes that crown the Valley of Mexico. The Iztaccíhuatl: the white woman; and the Popocatépetl: hill that smokes.

The snow waters of these gigantic volcanoes pass through vast layers of sand, running underneath the entire mountain range to reach the base of the mountains giving rise to twenty springs that form the lake of Xochimilco with its crystalline and pure waters. The most important of these springs was a hundred meters in width, ten deep, and its waters were sprouting from crystal rock, which drew all the objects with the colors and reverberations of the iris, a unique spectacle in the world.

The waters of all these springs ran into thousands and thousands of channels, which surrounded small islands, beautifully cultivated, called chinampas.[53]

Cuauhtémoc: Reconnecting Earth to Sky

The walk along the esplanade ventures back in time to find Cuauhtémoc (kwau-TEH-mock),[54] the last Aztec leader, fist perpetually to the sky.[55] His Nahuatl name means "descending eagle," and he wears a crown of feathers in his headdress. Upon a pedestal, facing east, he marks the center of the garden. »To each side of the figure, atop a set of towering narrow columns, an eagle with outspread wings perches on a cactus.« The eagles recall the legendary eagle that long ago perched on a rock in the middle of a lake, signaling Tenochtitlán, the ancient Ciudad de México, as the homeland of the Aztec people. In Urrutia's time, in a Mexico transformed by conquering forces, and to this day, this eagle remains on the Mexican flag as its most recognized national symbol. In one broad stroke, Urrutia references the seed of his existence and the first scene in the founding of a nation—eagles on a lake at the foot of volcanoes.

In 1921, Urrutia commissioned Luis L. Sanchez, a young Mexican artist, to create the statue of Cuauhtémoc, the first statue in the garden. Sanchez had excellent artisanal skills and experience creating cemetery statuary in Monterrey, Mexico. Urrutia worked closely in San Antonio with the sculptor, first providing him with a photograph of a Mexican rendition of the subject. Although Urrutia wanted an exact copy of the Mexican work, he soon found that his photograph did not contain enough information for Sanchez, and the two set out in collaboration, Sanchez with his creative and artisanal skills, and Urrutia with his scientific understanding of human anatomy.[56]

On a larger-than-life scale, the concrete statue of the renowned Aztec ruler crouches with all his energy thrust toward the heavens.[57] His gesture cries of defiance and protest, forming a sacred connection between earth and sky, reflecting the strength of the man who withstood capture and torture at the hands of Hernán Cortés in 1521 during the

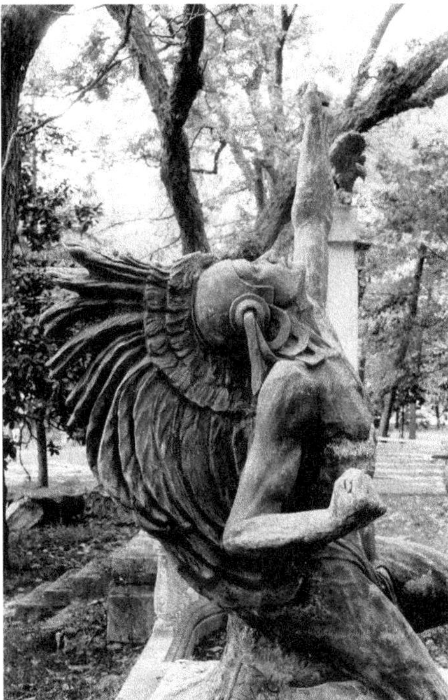

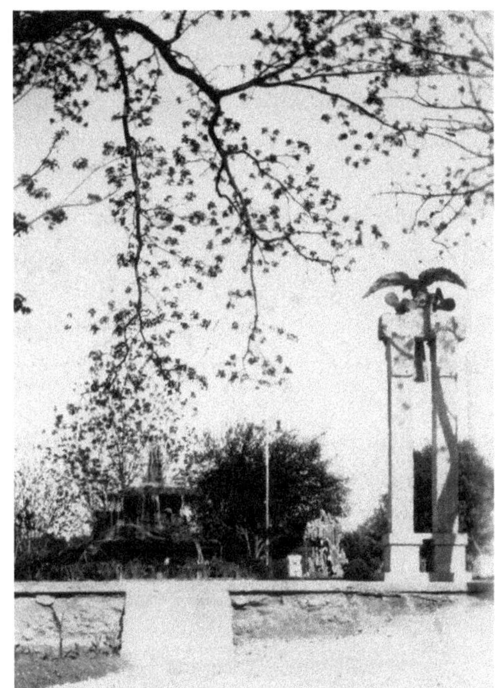

Left: Cuauhtémoc, 6 x 3 x 3 feet, excluding base, now in disrepair.

Right: View of esplanade from rose courtyard. Eagle (nonextant) on columns (now in ruins), concrete, 14 x 4 x 2 feet, including base. There were two sets of this design.

conquistador's quest to learn where Mexico's riches were hidden. Cortés was impressed with Cuauhtémoc's resistance and released him from capture. But four years later, overcome with paranoia, the Spaniard would forget his mercy and execute Cuauhtémoc, Mexico's last indigenous emperor, as the Aztec Empire crumbled under Spanish conquest.

As a young man, Urrutia was certainly aware of Mexico City's classic statue of Cuauhtémoc, created by sculptor Miguel Noreña. It was unveiled in 1887 not far from the Bosque de Chapultepec, towering high above the Avenida de la Reforma atop the monument celebrating the presidency of Porfirio Díaz, a man of both indigenous and Spanish heritage.[58] At the time, the Noreña sculpture was considered a bold representation of indigenous culture.

But Urrutia wanted a different representation of Cuauhtémoc, modeled after one he had seen at the Academia de San Carlos. Like Mexico's more public representation of Cuauhtémoc, the San Carlos version wears an elaborate feather headdress. However, in contrast to the more stoic Reforma statue, which is dressed in robes, the San Carlos

Cuauhtémoc wears only a loincloth and sandals, emphasizing his physical beauty and strength. And unlike the towering monument, the San Carlos version does not hold a spear or arrow with a forward gaze, instead directing his energy, anger, and defiance toward the heavens. Urrutia, in his work with the talented Sanchez, strove to embody this visceral Cuauhtémoc, connecting the forces of earth and sky at Miraflores.

To this day, Cuauhtémoc is an unofficial but powerful national symbol of Mexico, and Urrutia's choice of a relatively more expressive representation may be part of a larger trend of renewed pride in Aztec culture and history. Less understood outside the borders of Mexico, the Miraflores statue emphasizes the survival of the Aztecs in the four hundred years since Cuauhtémoc's encounter with Cortés but also, perhaps, seeks to reignite the story of indigenous Mexico in San Antonio's Mexican and Mexican American communities. Even with the decimation of the Aztec tribes, Cuauhtémoc lives on at Miraflores. Urrutia himself took great pride in his indigenous heritage.

At the statue's unveiling in December 1921, the well-known writer and exiled journalist Nemesio García Naranjo gave an apt analysis of the work as representing the outrage of Mexico's indigenous people. García Naranjo wrote:

> The entire statue, from the highest feathers of the plume to the clenched toes, cries of injustice, one of the major injustices in the history of humanity. Just remember the opulent constructions of Palenque, of Chichen Itza, of Mitla and of Teotihuacan to comprehend that something grand and divine was destroyed. And even though the colonial monuments proclaim the glory of one superior civilization, never will they constitute a convincing argument for the tragic sacrifice of a race. . . . For this the marble Indian protests: his clenched fist is that of Cuauhtémoc rebuking the gods and threatening the sky. . . . For this Urrutia has done well to place the touching marble on the shore of a lake: the pure waters silently collect the protest, for maybe tomorrow a protesting cloud will rise to the sky, that will burst into a storm of the rays of justice.[59]

Mexican Artists and Writers at Miraflores

Urrutia was one of San Antonio's early art collectors. As a wealthy physician circulating within the upper echelons of Mexico City society, he began collecting paintings and other valuable decorative objects around the turn of the century. The Mexican tradition of producing fine art at the time followed the European academic tradition of younger artists creating works that essentially duplicated that of the masters; hence, Urrutia had a number of recognizable copies of famous works in his collection. He brought some of his collection to San Antonio between 1914 and 1916. For example, he displayed a full-size copy of Rogier van der Weyden's *Descent from the Cross* in the waiting area of Clínica Urrutia.

At Miraflores, Urrutia's collection took on another layer of meaning. By making a garden filled with objects and landscape features recalling his homeland, he essentially created a metaphorical representation of the history and culture of central Mexico. Mexican elements such as natural landscape, indigenous gardens, Talavera tilework, classical sculpture, and Mexican architecture all combined to bring aspects of Mexico to *el México de afuera*.

For the landscape designs, Urrutia relied on his familial relationship with the land of Mexico, his intimate connection to Xochimilco, and his cultural and scientific knowledge of Mexican native plants. He also drew on his indigenous heritage as represented in folklore, oral tradition, and the Museo Nacional's early collection of Aztec relics, as well as museum paintings and personal experiences of the surrounding area's Aztec ruins and archaeological sites. The colonial Spanish architecture of the Ex-Convento de Churubusco and other buildings in Mexico City and Puebla, which incorporated Talavera objects and tile, inspired him, as did the sculptures in the Academia de San Carlos.

Urrutia, Dr. Atl, and La Academia de San Carlos

Urrutia's display of fine art at Miraflores, such as a replica of the Nike of Samothrace, discussed later, and Luis Sanchez's Cuauhtémoc, reflect his connection to the Academia de San Carlos, the first art school of the Americas, formed in the mid-1700s in Mexico City. Versions of these two sculptures were present in the academy's courtyard situated in the city's historic central area. The institution was located close to Urrutia's first medical office. His connection to the academy was due in part to his friendship with the artist Gerardo Murillo, who later took the name Dr. Atl. Like Urrutia, Murillo had early support from President Porfirio Díaz. Murillo began his studies at the academy in 1896, shortly after his arrival in Mexico City from his birthplace of Guadalajara. He not only excelled as a painter but as a world traveler, public muralist, curator, arts journalist, government arts representative, and teacher, ultimately becoming one of the first generation of Mexican modern art historians.[60]

Although he was a controversial figure who ultimately ventured far from the classicism of the academy, Murillo served as a professor, curator, and director there. He also was a patient of Urrutia's and attended his multidisciplinary salons in Coyoacán. After Urrutia's exile, the two friends corresponded. In 1929, as Urrutia contemplated an invitation from President Emilio Portes Gil to return to Mexico, Murillo, now Dr. Atl, proclaimed that he would be standing at the border to welcome him onto Mexican soil. It is presumed that Dr. Atl advised Urrutia on his art collection, and Dr. Atl's interest in Mexican landscape may have inspired some of Urrutia's ideas for Miraflores, such as the volcanic-style Fuente Monumental he commissioned from artist Dionicio Rodríguez.[61]

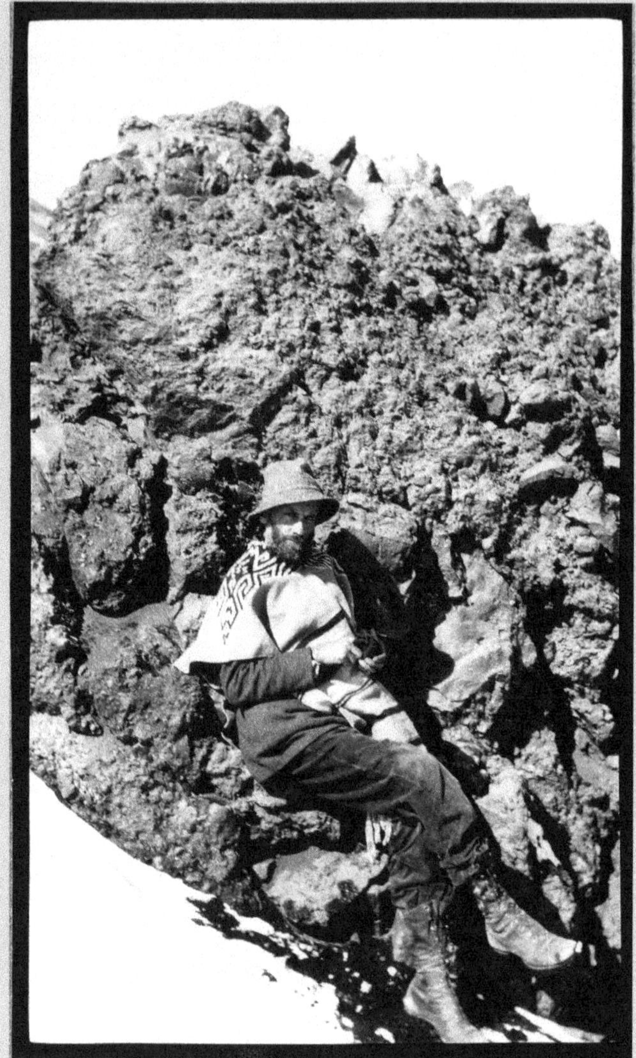

Dr. Atl, Popocatepetl, Mexico, 1921.

Urrutia may have been the first person to directly connect Mexican artists to San Antonio in the twentieth century, certainly preceding the country's rising interest in Mexican art that followed (for instance, the arrival of Diego Rivera to New York in the early 1930s). Sponsoring Mexican artists in the United States allowed Urrutia to extend his collecting activities into his life in exile, bringing more aspects of Mexico to San Antonio. It also gave Mexican artists a taste of the new life Urrutia was experiencing. And it exposed Anglos and Mexican Americans to Mexico at a closer vantage point than ever before, at a time when leisure travel to Mexico was relatively rare.

Luis L. Sanchez, a Young Artist's Start

Luis Sanchez Lopez, also known as L. L. Sanchez and Sanchez Lopez, the first artist to create a major work for Urrutia's garden, arrived in San Antonio around 1920. The young man lived on Dolorosa Street with his widowed mother, Antonia Lopez. The street was important to the Mexican immigrant community, home to residences and businesses, including a dough mill and offices for the publication *El Imparcial de Texas*. Porfirio Treviño, the Mexican architect of Urrutia's home on Broadway, may have occupied the space next door to Sanchez.[62] Sanchez was born in 1894 in Monterrey, Mexico. Treviño and Dionicio Rodríguez also lived in Monterrey for a time before coming to San Antonio. Perhaps Treviño, arriving in San Antonio in 1918, was the connection between Urrutia and these two artists.

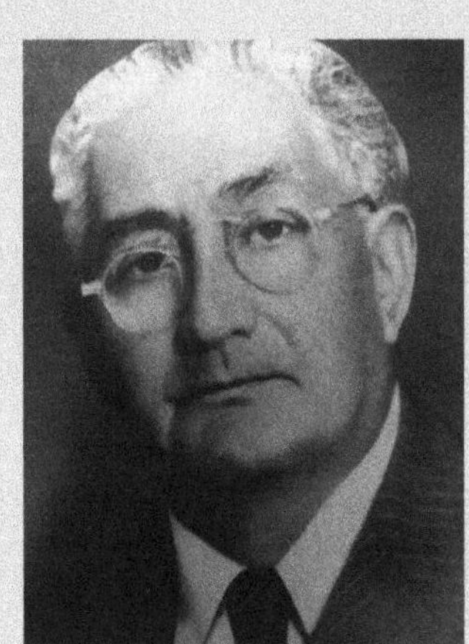

Luis L. Sanchez, aka Luis Sanchez Lopez.

Sanchez generally worked in concrete, with a traditional approach to constructing sculptural forms. He began his art education and training at age thirteen with the Decanini family, whom he started working for after the death of his father, Rafael.[63] Antonio Decanini (1878–1948) was an Italian engineer and sculptor in Monterrey, Mexico, where he and his wife are known for their creation of statuary for the city's Dolores and El Carmen cemeteries.[64] In San Antonio, Sanchez worked for the Rodríguez brothers, who established their business in 1920 and were popular manufacturers of monuments and statuary in central Texas.[65] According to Sylvia Sanchez-Pogson, granddaughter of Sanchez, when Urrutia hired Rodríguez Bros. to make several objects for his house garden at Quinta Urrutia, Sanchez's work, including the head of Coyolxauhqui (koy-yol-SHAUH-key) and the twin entrance lions, impressed him. In 1921, he commissioned Sanchez to create Cuauhtémoc as the first artwork for Miraflores, and later, the molded concrete lintel for the Urrutia arch, which was completed in 1924.

When he first came to San Antonio, Sanchez used the name Luis Sanchez Lopez, following the traditional Latin American custom of placing the paternal family name first and the maternal name second. As was true for many immigrant families from Mexico, however, this naming system was not accepted in mainstream life in the United States, and eventually Sanchez placed his father's name last, becoming Luis Lopez Sanchez apparently in order to conform to Anglo naming customs while still keeping his mother's family name, Lopez.

Sanchez's signature on two sculptures at Miraflores shows this evolution. The head of Coyolxauhqui, the mythical moon goddess of Aztec folklore, bears the signature "Sanchez Lopez." The statue of Cuauhtémoc bears the signature L. L. Sanchez, reflecting the transposition of his family names.

The Coyolxauhqui sculpture was later moved to Miraflores, and some years before Quinta Urrutia was demolished in 1962, Sanchez's twin lions were moved to Miraflores together with the replica of Nike, the Winged Victory of Samothrace.

In 1928 Sanchez moved to Harlingen, a Texas border town about thirty miles northwest of Brownsville, where he created an important public landmark. When South Ward elementary school was built at 309 W. Lincoln Street, a contractor commissioned him for some concrete artwork on the facade. Using concrete molds that he designed, Sanchez created an intricate scene of Quetzalcoatl with snakes on the front entrance doors and complementary designs around the nearby windows. Sanchez reportedly also created the special pigments for the design.[66]

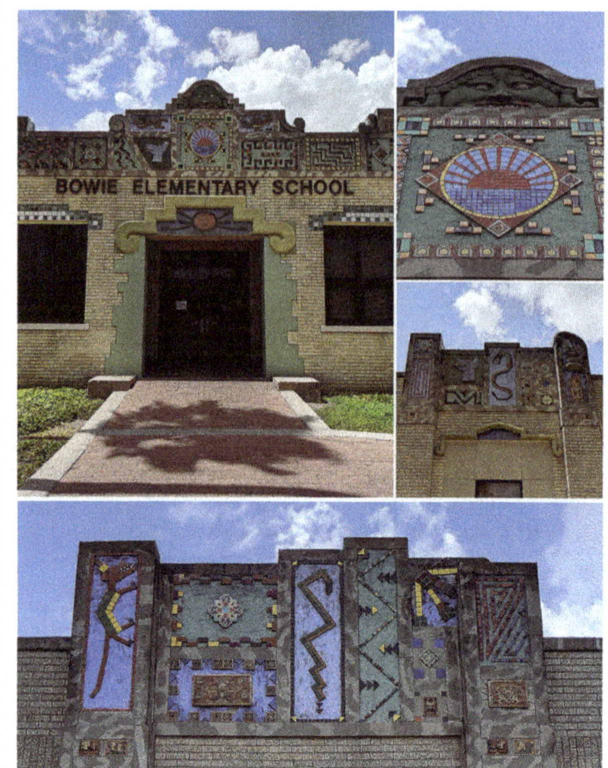

La Escuela de las Víboras.

Because of its vibrant design, inspired by Sanchez's roots in Aztec culture, the school became known as La Escuela de las Víboras (School of the Serpents). The features remain an inspiration to schoolchildren in the area and continue to convey the symbolism of the feathered serpent, Quetzalcoatl, as the messenger of wisdom and knowledge. The school was renamed James Bowie Elementary in 1935, but the Aztec decor and school's nickname are a respectful reminder of the multiple layers of heritage in this border region. The school recognizes and discusses the symbolism of Sanchez's work with its students, and connects its positive cultural message with the school's designation as a top Texas school, "the only Harlingen school to achieve the honor," from 2007 to 2012.[67]

Sanchez's work, which is likely the oldest existing public artwork in Harlingen, predates even the public murals of the 1930s Works Progress Administration. The artwork at La Escuela de las Víboras ranks highly among the city's many murals, which have become a well-known attraction for both visitors and community members.

Sanchez and his wife, Rosa Martinez, raised nine children. Their descendants still live in the Harlingen area. Sanchez died in 1958 at age sixty-two.

Marcelo Izaguirre, Architect and Engineer of the Gate

Marcelo Izaguirre was born in Culiacán, Sinaloa, in 1884. Marcelo and his wife, Esther, moved to Mexico City to establish his engineering career. His younger brother, Baltazar Izaguirre Rojo, was a noted poet and physician, and a friend and student of Urrutia's. Like Urrutia, Baltazar attended medical school at the Universidad Nacional de México in Mexico City. Throughout his accomplished career as a researcher in the field of tuberculosis, he maintained a successful literary career, publishing poems, poetry collections, and other writings.[68]

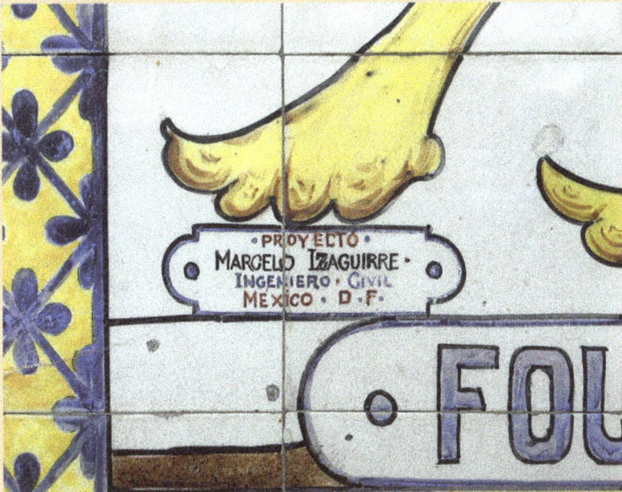

Marcelo Izaguirre's signature on Monumento a la Ciudad de México.

In 1921, twenty-seven-year-old *ingeniero civil* Marcelo designed the Monumento a la Ciudad de México gate for Miraflores, including the tile murals on each side of its two towers. Inspired by tilework at the Ex-Convento de Churubusco, the Talavera murals bear his name and celebrate the fourth centennial of Mexico City's founding and the founding of Miraflores in 1921. He most likely also designed the other buildings at Miraflores, including La Torre del Califa and Quinta Maria.

A great-grandson of the artist recalls that his *bisabuelo* was involved in the construction of the original offices for the Nacional Financiera, which was located near today's Torre Latinoamericana building. Marcelo Izaguirre died in Mexico City in 1940.

Dionicio Rodríguez, San Antonio's *Trabajo Rústico* Artist

Dionicio León Rodríguez was born in 1891 in Toluca de Lerdo, the State of Mexico's capital city, about forty miles west of Mexico City. In recognizing Rodríguez's remarkable talent for creating objects of a whimsical and yet convincing nature, Urrutia commissioned him to create a dozen or more objects, mostly in situ, to help articulate his vision for Miraflores. These were the artist's first works in the United States.[69] Although the work is sometimes called faux bois, which describes an artisan technique in the European tromp l'oeil tradition of mimicking wood using other materials, the similar Mexican tradition is called *trabajo rústico* and imitates not only wood but also stone and other objects using special mixtures of concrete. Even today, a number of Mexico City's parks teem with *trabajo rústico* bridges, waterways, fountains, railings, benches, and more.

At the turn of the century, Rodríguez created enormous boulders in and around the lake and its fountain in Mexico City's Chapultepec Park, a place certainly familiar to Urrutia. Miraflores contained one of the

Trabajo Rústico Collection

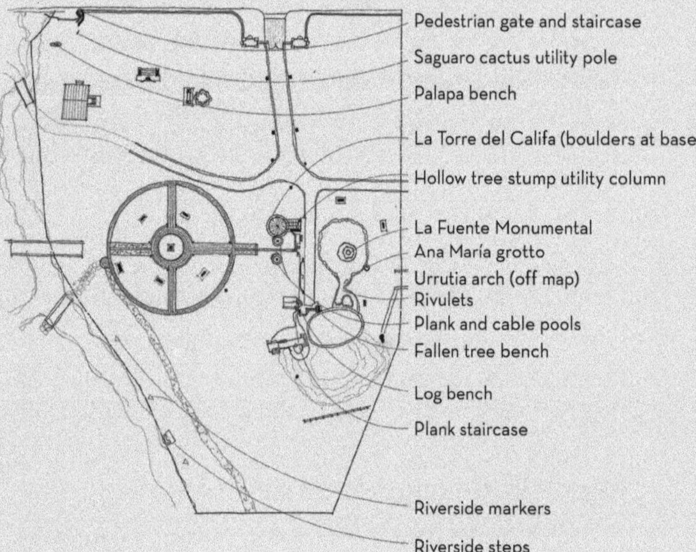

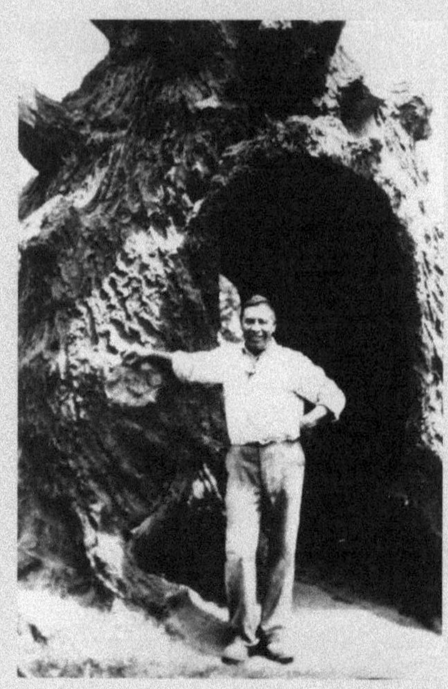

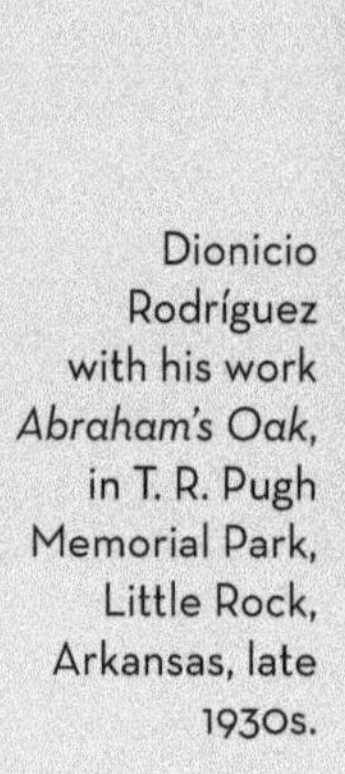
Dionicio Rodríguez with his work *Abraham's Oak*, in T. R. Pugh Memorial Park, Little Rock, Arkansas, late 1930s.

largest and most varied collections of Rodríguez's work in one setting, and the artist's career ultimately took him from San Antonio to other locations throughout the South and Midwest for nearly three decades.[70] Rodríguez is said to have been highly secretive about the processes and formulas for creating his medium, and little is known about the pigments he used.[71] Even today his proprietary concrete mixture defies duplication.

Urrutia was undoubtedly a long-term collector and a primary advocate for the artist's work. Rodríguez's first creation in San Antonio is thought to be the faux stonework in the Urrutia arch, which was completed by 1924. Urrutia expressed his enthusiasm for it in a letter to Ray Lambert, San Antonio's commissioner of parks. Brackenridge Park nearby became a public site for Rodríguez's work in San Antonio. His remarkable pedestrian bridge, built in 1925, sits in the park across the San Antonio River from Miraflores.

At Miraflores, Rodríguez went on to mimic many forms of wood (hewn planks, cut logs, natural tree branches and trunks), as well as numerous types of stone (volcanic, limestone, hewn, natural, cave formations), rope, thatched roofs, plants, and even iron. Rarely solely sculptural in nature, the objects usually functioned as benches, fountains and other water works, staircases, gates, bridges, and housing for electrical equipment. As with the Talavera seats, Urrutia used Rodríguez's works in almost every sector of the garden, making them an enchanting and unifying element. They had the added impact of playfully reminding visitors that all is not necessarily as it appears at first glance, a tenet Urrutia greatly appreciated.

Rodríguez's works in cement are long-lived, easily lasting over a century, though some have been destroyed and many are now in need of restoration. »Among Rodríguez's most impressive and unique works was the faux stone Fuente Monumental fountain, which was disassembled in 2001.«

Rodríguez is said to have preferred solitude to companionship. He had one brief, failed marriage, suffered from diabetes, and lived to the age of sixty-four, dying in 1955. For the last twenty years of his life, when he was not traveling to install his commissions, he lived alone in a *trabajo rústico* tree-and-cave-like structure that he built on a plot of land on Guadalupe Street near the San Fernando Cemetery where he now rests.[72]

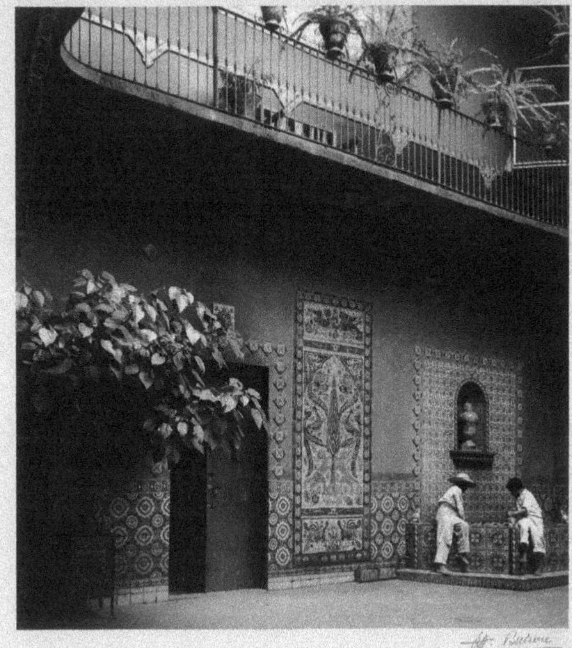

Main patio at Uriarte Talavera, Puebla, Mexico, 1933.

Ysauro Uriarte Martínez, Talavera Ceramicist of Puebla

Several signatures on the tiles of the Urrutia arch proclaim the origins of the colored *azulejos* that rustically decorate its surface. At the foot of the left column mural—the cypress tree flanked by peacocks at the top and jaguars at the bottom—rests the signature "I. Uriarte." The lintel bears several signatures including the decorative "Y.U." in two places, along with the identifying location "Puebla-Mex."

The Uriarte workshop was founded by Ignacio Uriarte in Puebla around 1824. The clay-rich town of Puebla became the capital of the Talavera pottery industry established by the Spanish in the 1500s. Over the next two centuries Talavera workshops populated the area, and the colorful, intricate, interior and exterior tiles they produced are still seen throughout the town and in Mexico City.

Although a well-regulated Talavera industry flourished in Mexico from 1650 to 1750, with about thirty workshops producing the pottery, by the first half of the nineteenth century the industry was well into a decline. Having abandoned the stringent production guidelines, and with a scant number of manufacturers, the industry remained depressed until the turn of the twentieth century, at which time Ignacio's workshop had passed along to his son, Dimas Uriarte. Enrique Luis Ventosa (1868, Barcelona–1935, Puebla), educated in Paris and a scholar of Spanish Talavera, had come to Mexico to study and revive the Mexican Talavera process.[73] In 1922 he partnered with Ysauro Uriarte, Ignacio's grandson (born in Puebla, 1884), who had inherited the Uriarte workshop from Dimas. The two men restarted the Talavera industry with an authentic level of quality and technique.

The origins of Urrutia's Talavera collection have not yet been fully explored, but one can observe the difference in the quality of the tiles on the Urrutia arch lintel, which must have been made months or even years before 1924 (the year the arch was completed), and the similarly designed tiles, displayed in the McNay Art Museum courtyard, which were installed around 1927. Although the Urrutia mural has lacked the attentive care that the McNay Art Museum bestows on its outdoor Talavera mural, the Urrutia tiles can be presumed to have a more rustic appearance because they date from pre-1922, when Talavera production was of a lower quality. In any case, it is clear that the Urrutia tiles originate from the Uriarte workshop and represent part of the history of that tradition.

Ysauro Uriarte died in Puebla, Mexico, around 1971 at age eighty-six, but the house of Uriarte is still in production today.

Pantaleón Panduro, a Link to Ancient Ceramics

Pantaleón Panduro Martínez (1847–1909) lived in Tlaquepaque (t-la-keh-PAH-keh, the "place above clay" in Nahuatl), a town outside of Guadalajara, where he became known as one of Mexico's best ceramicists of the nineteenth century. In 1890, the Museo de Productos Industriales collected Panduro's ceramic sculptural work as part of its effort to promote Mexico's industrial excellence. Housed at the Engineering School at the College of San Juan in Guadalajara, the collection included sculptural ceramic busts of various illustrious Mexicans. Panduro's sculptures were appreciated for their "remarkable resemblance" and were "well known in the country and abroad."[74]

The 1892 Exposición Histórico-Americana, a conference in Madrid exploring the pre- to post-Columbian era

406 SCIENCE. [Vol. VIII., No. 196

1. — Pantaleón Panduro, the Guadalajara potter.
2. — His clay board, showing the raw material.
3. — Appearance of his paste when worked up.
4. — Spatula for cutting and smoothing.
5. — Furnace of tiles cemented with clay.
6. — Trowels and decorating-tools of iron or rosewood.
7. — Brushes for painting, bristles of yucca fibre.
8. — Burnisher of hematite set in a clay handle.
9. — Scraper of tin.

Illustration of Pantaleón Panduro and his tools.

in the fourth centennial celebration of New Spain, recognized Panduro as "the most skillful of the Guadalajara figurine makers . . . adept at modeling from life, and his figures and groups are much sought after." One writer noted that Panduro's work related to pre-Columbian clay processes used in the generations before, including vessels and the "small figurines" characteristic of the area. The writer observed that the Museo Nacional held excellent examples of this type of ceramic work.[75]

A few years later, the Mexico exhibit at the 1886 New Orleans World's Fair brought further attention to the industries of Mexico, and the United States National Museum (which eventually became the Smithsonian Institution) took interest in Panduro, collecting samples of his clay and tools and documenting his kiln and various stages of his work in progress. Scholars were in search of information about the history of ceramics in Mexico, and they considered Panduro to be a link to ancient Aztec and Mayan methods, particularly his use of an "open furnace, in which the ware can be hardened but not glazed."[76]

In 1900, Owen Wallace Gillpatrick, an American writer and avid traveler in Mexico, wrote a profile of Panduro for *Brick*, a magazine focused on the ceramic industry. He reported:

> I have just come from the workshops of those remarkable artists, Pantaleón Panduro and his son Timoteo. The older man has the head of a lion, but it is that of a very kind lion, with the benevolent look that I somehow feel should always go with genius. . . . Panduro the elder modeled in clay long before he began making portrait busts, for which he is now famous. His first experience was unique. A certain bishop, who was greatly beloved by all, had died, and one of his closest friends came to Panduro and requested him to make a cast of the dead man's features. The artist declared that he was not competent. The friend insisted, and he finally made the attempt. The result was a wonderful likeness of the bishop, and the beginning of Panduro's career.

Panduro's son Timoteo, who took up his father's profession, told Gillpatrick that "the crowning event of his life was his visit to Mexico [City] with his father, to make a portrait of President Díaz. As he speaks of this, his voice takes on a reverential tone, and it is easy to see that his president is his idol."

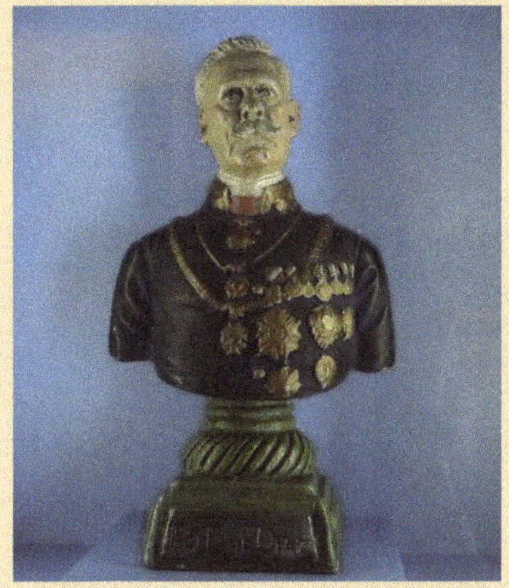

Bust of Porfirio Díaz, after Panduro. Museo Nacional de las Intervenciones, Mexico City.

»Whether the bust of Porfirio Díaz at Miraflores was an original or a faithful copy is unknown, but Panduro's renowned representation of Díaz ranks as one of his best-known works and has been reproduced extensively. The sculpture of the water carrier that once stood in the center of Urrutia's pond at Miraflores was also by or after Panduro. For Urrutia it recalled Guadalajara. Unfortunately, both statues no longer exist at Miraflores, and their whereabouts remain a mystery.« Many descendants and followers of Panduro continue his ceramics tradition, and a museum in his name honors his work and that of Jalisco potters worldwide.

Ignacio Asúnsolo, Mexico City's Revolutionary Sculptor

Not enough has been written about the important work of Ignacio Asúnsolo, a prolific sculptor who worked primarily in bronze and stone. But Margarita Nelken, in her 1962 book *Ignacio Asúnsolo* tells an essential story about the artist. Born the son of a powerful hacienda family in 1890 in Parral, Chihuahua, Asúnsolo moved to Mexico City to attend the Academia de San Carlos in 1910. One day he came to class wearing a red carnation, a symbol of revolution, which his professor, a follower of Porfirio Díaz, considered an affront. Asúnsolo was suspended for a week and lost his scholarship. He returned to the school but worked at the school of medicine to support himself while participating in the armed struggle to bring Francisco Madero to the presidency.

In 1913, Asúnsolo briefly ceased his political activities and returned to San Carlos, gaining an appointment as the chair of drawing. However, the Huerta presidency reignited his political activism, for which he was arrested, held for days, and beaten severely. According to Nelken, a relative, hearing of Asúnsolo's dire situation, approached the secretary of the interior, who intervened by sending José Juan Tablada with an order to release Asúnsolo from custody, saving his life. Friends from the medical school cared for Asúnsolo, who discontinued his demonstrations shortly thereafter.[77]

During the time Asúnsolo worked at the school, from 1911 to 1913, Urrutia was the chair of surgery. In January 1913 Urrutia was named director of the school, and he was Huerta's secretary of the interior from June to early September 1913. Urrutia was also a close friend of Tablada's and had connections to the Academia de San Carlos as an art collector. Tablada himself was a close friend of Asúnsolo's cousin Francisco Asúnsolo, a banker. Because of these connections, it is reasonable to assume that Urrutia was the secretary of the interior referenced by Nelken, who directed Tablada to rescue Asúnsolo from his 1913 arrest and to deliver him to medical students at the school to care for him.

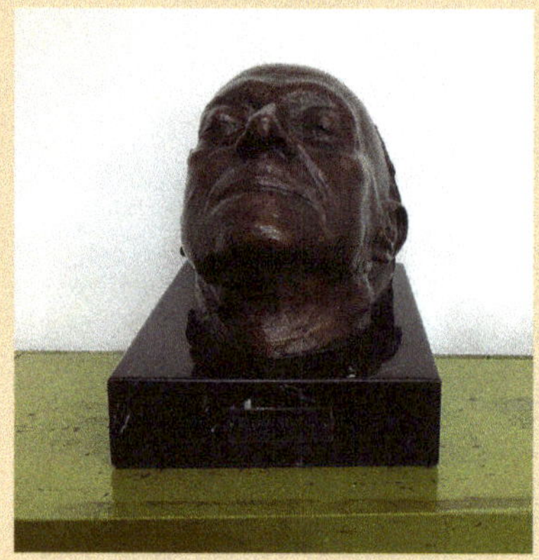

Máscara Mortuoria (Diego Rivera), Ignacio Asúnsolo, 1957.

Asúnsolo went on to study at the École des Beaux-Arts in Paris and created many sculptures, many of which capture the spirit of the people of Mexico. In addition to portraying Urrutia (1940), he also memorialized other notable Mexicans, including Sra. Juana Inés de la Cruz (1922), Álvaro Obregón (1934, in a collaborative public work), Justo Sierra (1945), Emiliano Zapata (1950), Luis Cabrera (1954), Francisco Villa (1957), Diego Rivera (1957), and Francisco Madero (1960).[78] Asúnsolo served as director of San Carlos from 1947 to 1954 and continued thereafter as a professor of sculpture. He died in 1965 in Mexico City.

José Juan Tablada, Modern Mexican Writer

Born in Coyoacán in 1871, José Juan Tablada was an important friend to Urrutia and participated in his salons at Sanatorio Urrutia. They met in Mexico in 1905 and suffered the common experience of exile beginning in 1914. Despite his exile, Tablada was a giant of Mexican literature, publishing prolifically, especially poetry books and magazine articles, and working in fiction, biography, art criticism, translation, and journalism. He settled in New York in 1920 but traveled extensively to Paris and Japan, and he was singularly responsible for introducing haiku into the Spanish language.[79] In Mexico he served in the leadership of La Escuela de Bellas Artes and as a professor of archaeology at the Museo Nacional. In New York he ran a bookstore that catered to the Latino community.[80] Of his early successful poem "Onix," Tablada wrote to Urrutia, "Aureliano, though I made these verses before I met you, keep the original because I am certain that your spirit ever since has determined to direct my life."[81]

Tablada saluted Urrutia as "*un hermano de corazón y de espíritu*" (a brother in heart and spirit), and the two men held each other in high regard. They corresponded over the years, especially from 1927 to 1931, during which time Urrutia commissioned Tablada to pen biographical sketches about Urrutia's life. The arrangement was handled with enthusiasm, sensitivity, and mutual admiration. For Tablada, the commissions were a break from the tedium of journalism that was barely supporting him financially, and for Urrutia the sketches shined a light on his life in Mexico, his accomplishments as a physician-surgeon, and his short though tumultuous experience in politics and government.

Tablada had an intricate understanding of Miraflores's significance and wrote about it in his poem "Nueces de Miraflores," a version of which he gifted to Urrutia and which utilizes the idea of a tree nut growing in the garden and the concepts of seed, tree, and fruit to transport one to and from specific places and ideas of central Mexico. The poem concludes with the idea that no matter where we live, the same sun, moon, and stars illuminate the heavens; Urrutia has succeeded in San Antonio just as he did in Mexico. The two men supported each other with their strong friendship, which helped to make life more bearable as they struggled to live away from the country they both loved.

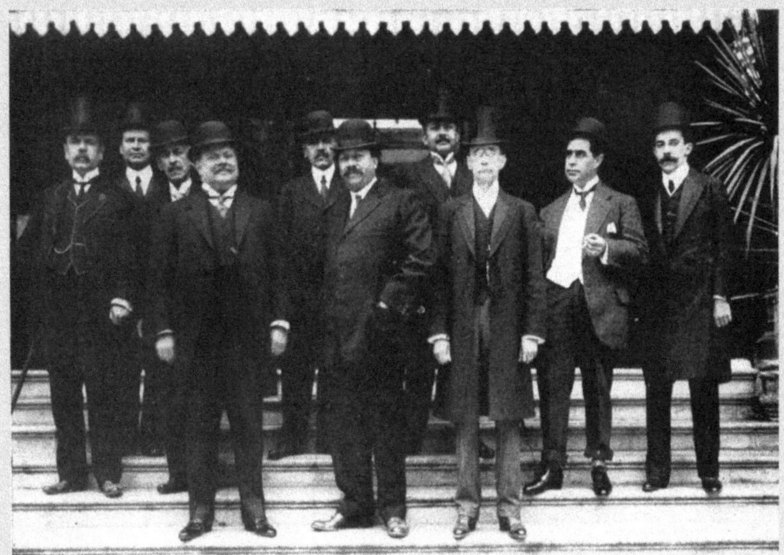

Urrutia, center, with colleagues from the Universidad Nacional, including José Juan Tablada, second from right, ca. 1913.

Due to Mexico's changing political landscape, Tablada was never able to resettle in his homeland (a problem Urrutia also recognized for himself), although early on the author was able to reconcile enough with the government to write for Mexican publications. He lived in Mexico briefly toward the end of his life and was named to La Academia Mexicana de la Lengua in 1941. After his death in 1945 in New York, his remains were moved to Mexico City. He is buried in the Panteón Civil de Dolores and he is included in the Rotonda de las Personas Ilustres.

The following are a few excerpted stanzas (in translation) from Tablada's poem "Nueces de Miraflores" that show the connection Tablada had with Urrutia and his garden:

***
Nueces de Miraflores, you are seeds
And so you sprout in my mind
Rumbling trees . . . remembrances
Of the memory in the deep jungle!
Noble trees under whose shadow
Prairies fill with yearning
nostalgic like wild flowers
The Sun opening its fresh petals.
Under those fronds and above the grass
They dance an angelic round
Nymphs and Muses—Ay!—the illusions
of our youth in Spring . . .

Is it dawn or twilight? . . . Golden
The old church emerges from Coyoacán;[82]
Volcanoes peaked with the whiteness
Of eternal snows;
The Pedregal[83] extends the swell
Of still and stony waves;
Churubusco[84] reawakens from other days
The tragic epic;
Sapphire and lapis lazuli withdraw
And further, behind the sandy banks,
Del Ajusco,[85] the brave mountain range

Oasis at the foot of the jagged escarpment,
For her rustic furrowed gondolas
From Anahuac[86] to Venice;
Eden for romantic idylls;
Of the lacustrine Empire's last pearl;
Chalchihuite[87] among emerald feathers
And the Valley of Mexico[88] the gem
All the blue of Xochimilco's[89]sky
Collects in her turquoise lagoons!
***
Nueces de Miraflores, destiny
Sprouted you in San Antonio de Béjar
When in the hand that sowed you, the pure
Ancient blood runs through the veins,
When the providence of the planter
Is identical in his zeal
To that which formed the chinampas
With fistfuls of earth
Like the nests of the swallows
Industrious, cheerful potters . . .

Well, but for hostile Destiny
The same thoughtful hand would have
There in the heart of Xochimilco
Sprouted your beauty better
Either giving shade to the bustling market
Or to the familiar plaza in front of the church!
***
But what does it matter? . . . What are your leaves
Toward the blue our dream rises,
God is one, there is no fatherland in heaven
Ours is the Sun, the Moon, and the Stars
The same that illuminate the Homeland
The volcanoes, the trees, the forests.
The same divine light that faithfully portrays
Xochimilco in its turquoise lakes . . .
***[90]

Nemesio García Naranjo, Writer of Exile

Nemesio García Naranjo, born in 1883, made his way to Mexico City from the small town of Lampazos, north of Monterrey, to attend law school. A supporter of Porfirio Díaz, García Naranjo received his law degree in 1909. He served in the legislature under Díaz, and in Huerta's cabinet as minister of education from 1913 to 1914. He was yet another 1914 exile and intellectual who attended Urrutia's salons at Coyoacán, and the two men were close friends.

In 1920, García Naranjo paid tribute to Urrutia at the celebration of his forty-eighth birthday, twenty-fifth year as a physician, and fifth year as a surgeon in San Antonio. Of Urrutia's abilities as a teacher, he said: "Science may be dry and small like a seed, but it must be dropped tenderly and lovingly in the furrows so that it can grow. This is the sower's secret."[91] He also said that Urrutia "has a deep knowledge of the natural sciences and is passionate about the fine arts. A painting of vivid colors moves his spirit, and a delicate melody softens his thoughts. That's why in Mexico he fraternized with artists, and here in exile he cultivates the friendship of poets."[92]

García Naranjo arrived in San Antonio a few months after Urrutia and founded *Revista Mexicana*, a publication dedicated to providing the exile community with news of the homeland and its counterrevolutionary voices.[93] He wrote about Urrutia and other subjects for both the *Revista* and the more successful Spanish-language newspaper *La Prensa*. His 1921 article about Urrutia's first installation at Miraflores, Luis Sanchez's statue of Cuauhtémoc, analyzes the unique aspects of Sanchez's expression as it reflects Urrutia's point of view as an exile. Familiar with Urrutia's former life in Mexico, García Naranjo sees that the sculpture, in its posture of protest, echoes Urrutia's pain even as it depicts the pain of injustices suffered by the Aztec nation.

García Naranjo eventually repatriated to Mexico. His extensive memoir navigates his illustrious life both in and outside the country. He was inducted into the Academia Mexicana de la Lengua in 1940, becoming one of his country's most highly decorated writers, and lived in Mexico City until his death in 1962.

Federico Allen Hinojosa, Journalist of *el México de afuera*

Born in 1888 in Lampazos, the same town as Nemesio García Naranjo, Federico Allen Hinojosa came to the United States as early as 1912. He was a career journalist, writing for a number of Spanish-language publications and serving as a managing editor at *La Prensa*. Allen Hinojosa, García Naranjo, and Urrutia often participated alongside others in the exile community in social and charitable activities to strengthen San Antonio's cultural ties with Mexico and to address challenging immigrant and Mexican American issues such as education, employment, and health care.

In June 1944, as Urrutia approached his seventy-second birthday, Allen Hinojosa interviewed him about his perspective looking back on the Mexican Revolution and toward the future of Mexico. The interview took place at Miraflores, and was a rare occurrence, according to Allen Hinojosa. Urrutia appeared upbeat and hopeful about the future. In the previous decade, at the height of his medical career, he had been honored by the Brazilian government in 1935 as principal speaker at the sixth Pan-American Medical Congress and inducted into the American College of Surgeons in 1936. A few years later, he had received his US citizenship in 1942. In the interview he speaks passionately, comparing the revolution to a body that recovers more strongly after fighting off a life-threatening illness. He emphasizes themes of freedom, rights, and responsibility, anticipates the 1946 Mexican presidential election, and points out that 1947 will be the centennial of the Mexican-American War. The article captures the mature eloquence of both men, as Allen Hinojosa expertly weaves Miraflores's idyllic and refreshing setting with the doctor's personality:

> With the beautiful wrought-iron fence that forms the main gate of Miraflores open wide, we enter the magnificent garden that Dr. Urrutia cultivates with so much love, at the end of Brackenridge Park, between Broadway and Hildebrand. We are expectant. A strong mastiff lying at his feet gets up and starts to come toward us. Dr. Urrutia calls it and it returns to its sleep. While close by, we are sure that we have nothing to fear.
>
> Dr. Urrutia is alone, sitting on one of the many tiled benches in Miraflores, facing the monumental fountain, his back to the Tower of the Caliph. Twilight sets a splendor in the admirable landscape of this charming place. The eve has rained torrentially and while in other parts of the city it has not left the smallest trace due to the tremendous drought of several months, in Miraflores it seems that the rain has hardly stopped falling. Having passed the wide gate and entering under the shade of pecans, cypresses and alamos, we have stopped feeling the suffocating heat that surrounds the city. It seems that in an instant we have left San Antonio to enter a paradise.
>
> And look—steadily the high cypress trees swing gracefully touched by the breeze of dusk. Their canopies shine with the bright glories of the sun, like pearls that the rain of the previous night has left in them.[94]

Allen Hinojosa concludes:

> We are surprised when Dr. Urrutia pauses staring at the bottom of the pond, where there are some exquisite irises. The eye marvels at the miracle of color that emerges from the calm water. He seems distracted; perhaps he believes that we have not been attentive to his speech and that we have lost his meaning, in which he put energy and sincerity. But with that innate quickness of thought in him relying on his artistic expressive sense and his love of flowers, he says, putting his hand on my shoulder: "How beautiful, yes? Then you will be more impressed to know that these lilies are of those that Maximilian brought to Mexico. They are from the Nile, but Maximilian, who loved Mexico so much, wanted to bring the best for his new country. I brought some here, taking them from the oblivion and abandonment where they were in Cuernavaca, and here you have them. Perhaps there, where he had them planted, they no longer exist!"
>
> The doctor rises and invites me to take a walk along the paths of the garden. Then a refreshment is served which we savor as the splendor of the evening melts away.[95]

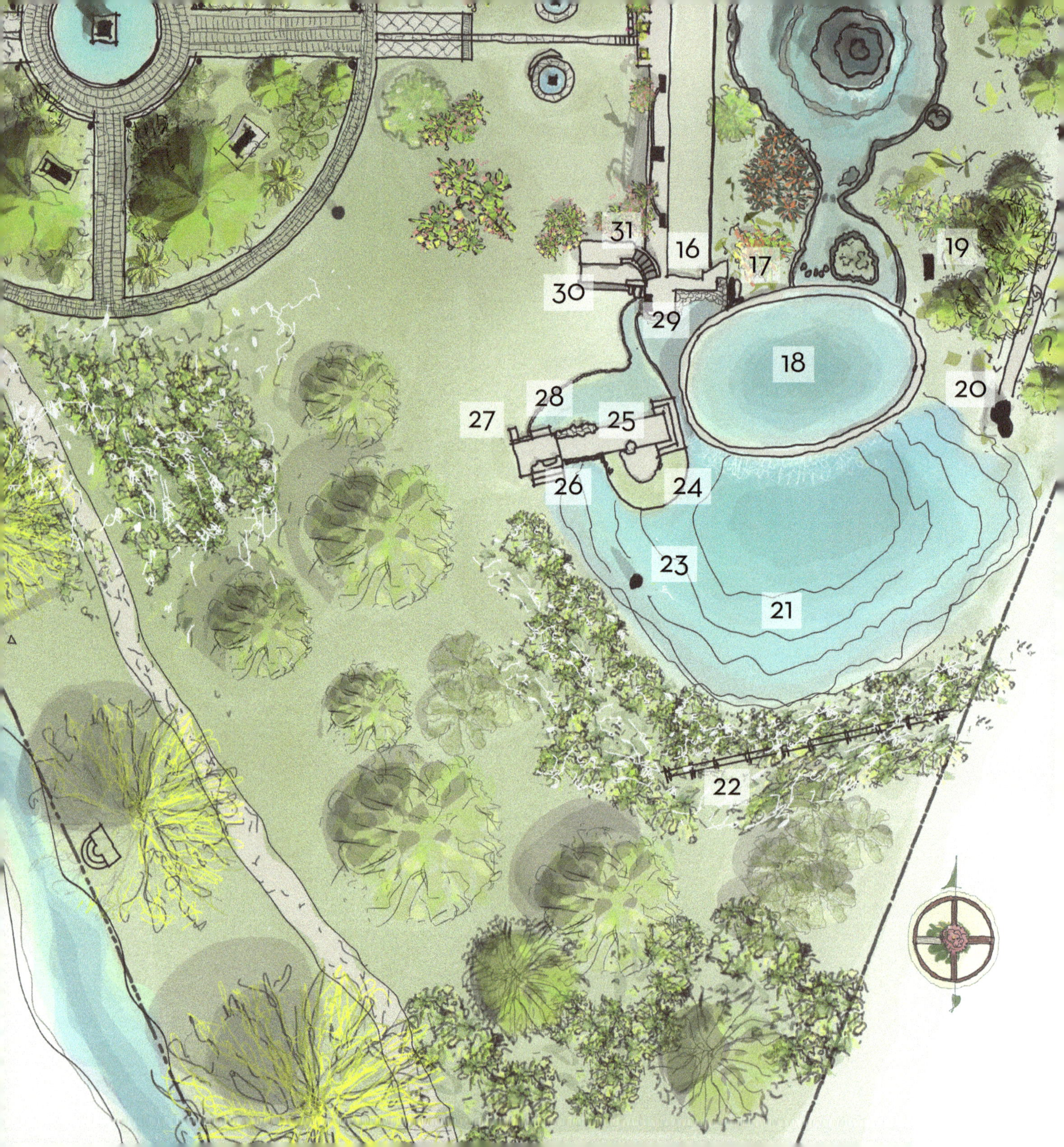
31
16
17
19
30
29
18
20
28
27
25
26
24
23
21
22

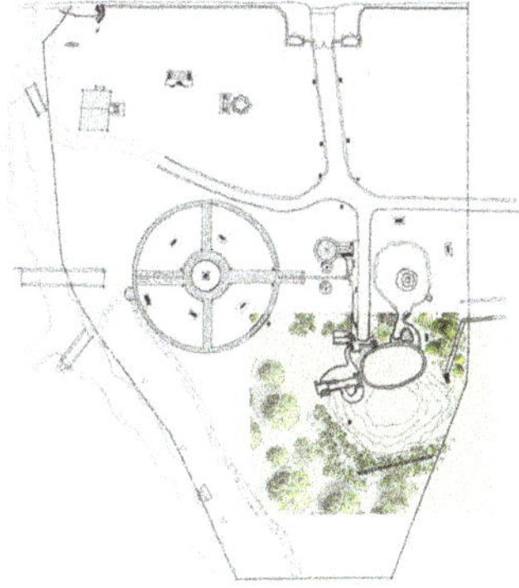

16 Viewing platform
17 Fallen tree bench
18 El Baño de Nezahualcoyotl
19 Jaguar bench
20 Coyolxauhqui
21 Pond
22 Arbor
23 Water carrier statue
24 La Isla Luz Fernández
25 Stage bridge
26 Two-sided bench
27 1716 stone
28 Step bench
29 Log bench
30 Stepped seating
31 Plank staircase

The Watery Garden of the Poet King

Beyond Cuauhtémoc, nature takes over as the esplanade drops sharply from its viewing platform. »The highest point of Miraflores gives way to views of wooded parkland to the south.[96] The waters of an egg-shaped pool, about sixty by forty feet, swell to its top, and its glassy surface drips and dribbles over the edge, overflowing into the beginnings of a larger pond below. The waters expand, in one direction lapping the rounded pool underside as it juts out from the esplanade wall, buttressed and exposed, and in another direction, travelling under a flat, broad bridge. The flow converges in a quiet pond occupying the southern edge of the property. The bridge stretches from the pond's west bank toward the bulbous pool wall, holding a series of lily ponds below the esplanade on one side and the larger pond on its other side. Cypress, willow, elephant ears, and other water-loving plants of subtropical Mexico border the pond, which expands and contracts with sparse or plentiful rains. Beyond it lies a wilderness of trees and tall grasses.«

Visitors may have arrived through the Spanish-style gate or even the hybrid-style Urrutia arch, but the surroundings at the top of this watery lagoon represent Urrutia's reverence for his indigenous heritage and the natural world. The area, which he calls El Baño de Nezahualcoyotl (the Bath of Nezahualcoyotl, pronounced nesa-hual-COY-otl), stands between the statues of Cuauhtémoc at one edge of the pool and the moon goddess Coyolxauhqui at the other, in the garden, and so too does it reflect the place of Nezahualcoyotl in the timeline of Aztec culture. With Coyolxauhqui representing the origin story and Cuauhtémoc the end of Aztec culture as it was, Nezahualcoyotl perhaps represents the culture's most prolific and sophisticated flowering along this ancient path.

El Baño de Nezahualcoyotl: Back to the Garden

Urrutia swims in the dark, seemingly bottomless pool almost daily, in the nude, even in the winter.[97] He often sits on the *trabajo rústico* fallen tree bench nested between the esplanade and the pool's edge, reading in the evenings until twilight.[98]

Perhaps, at that moment between night and day, he looks up from his book and out into the wilderness and dreams of another Mexican garden, created in the fifteenth century by the legendary Aztec ruler Nezahualcoyotl. In his ancient garden at Texcoco, located in what is now the northeastern corner of Mexico City, medicinal plants were grown, collected, studied, and used extensively. Hundreds of long-lived, shady ahuehuetes, or Montezuma bald cypress, reigning supremely above all Aztec trees, guarded the winding paths and arbors. Water flowed and the eagle and jaguar joined the garden with sky and sun, earth and night. The Aztec king of Texcoco "laid out his garden on the sacred hill . . . and used the canalized waters flowing down from the springs of Mount Tlaloc to create ponds with panoramic views from the hillside," which was reached by a curving path of rock stairs.[99]

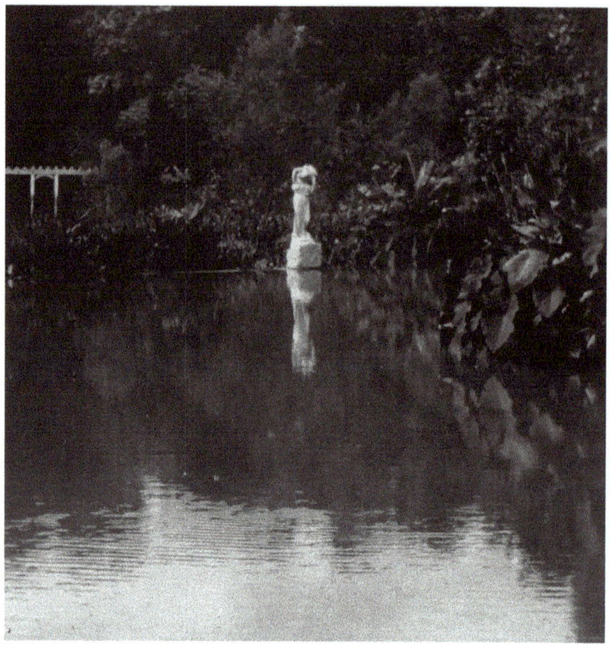

Pond with water carrier and arbor, all nonextant.

A memory, or not really a memory but the inherited remembrance of the king's garden, lives on. Urrutia draws on the tradition of cultivating medicinal plants. He grows them, studies their properties, uses them, and sees his patients benefit. »His ahuehuete grove descends from the garden seeds of Texcoco.[100] Here, too, the eagle and jaguar reign. The jaguar lurks nearby not only at the arch but also in a modest but classical pre-Hispanic representation on a concrete jaguar bench.[101] The eagles of Cuauhtémoc, perched atop the pillars flanking the last Aztec ruler, peer down upon the more ancient king's memory, attesting to the staying power of the once threatened people.« Urrutia's version of Nezahualcoyotl's garden even includes its own curving faux bois plank staircase to connect the levels of the garden.

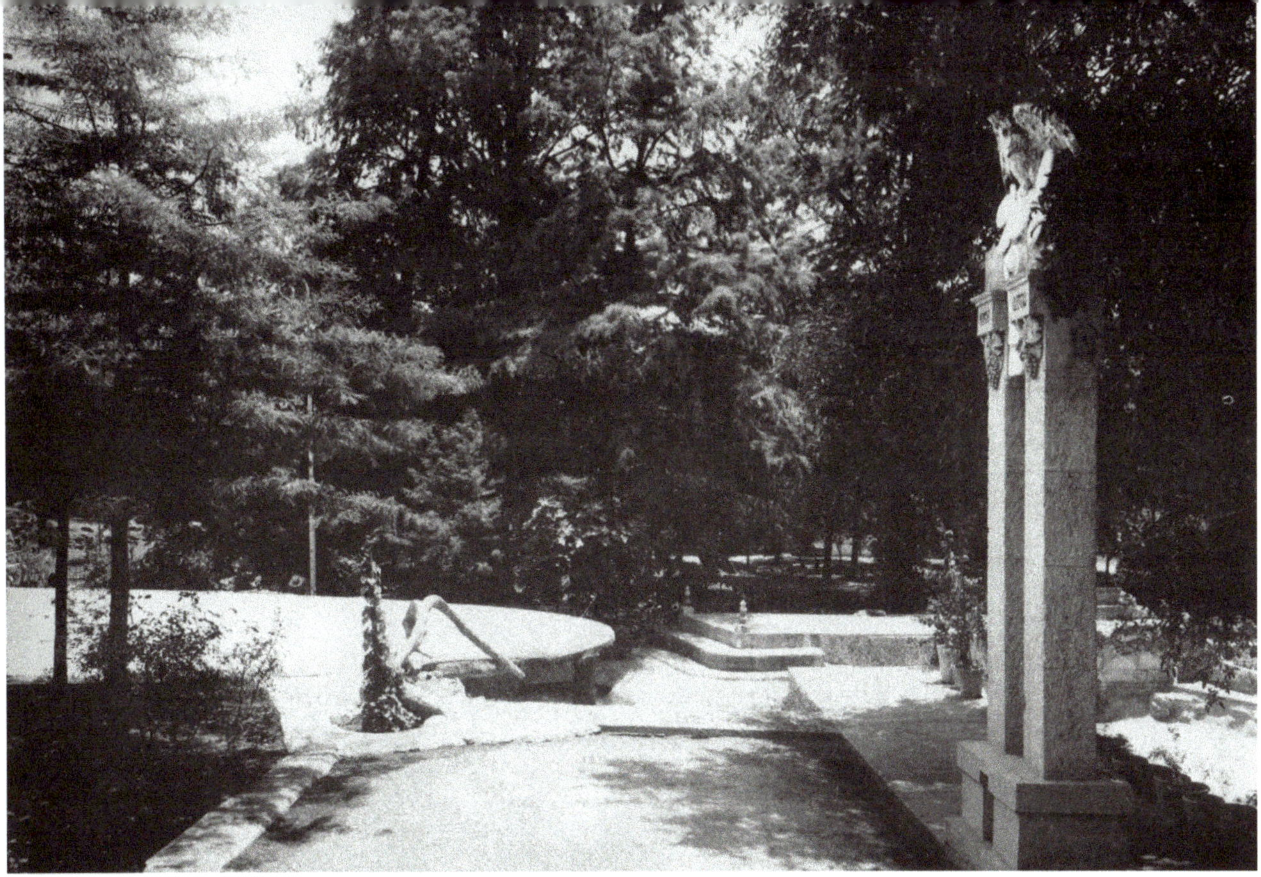

Baño de Nezahualcoyotl at left, concrete, 60 x 40 feet, nonextant.

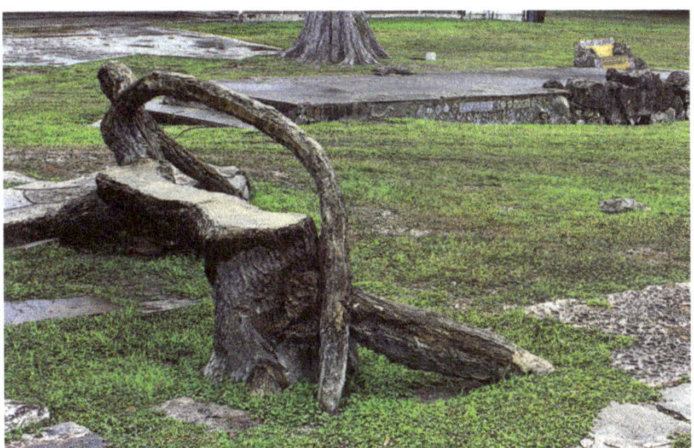

Fallen tree bench, *trabajo rústico*, 3 x 12 x 6 feet.

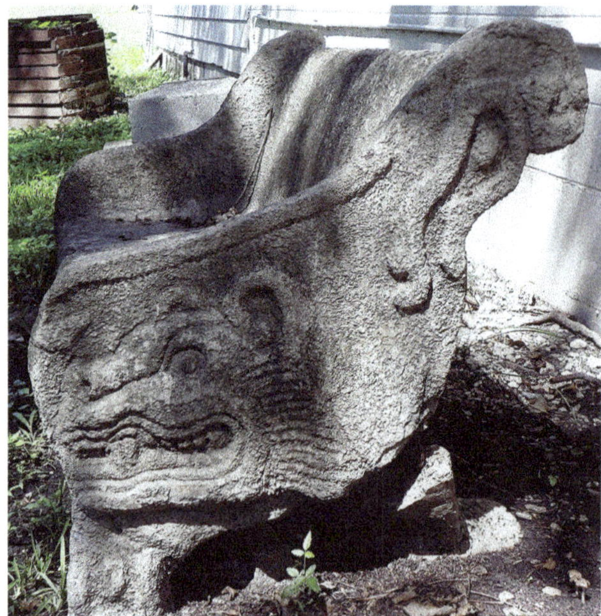

Jaguar bench, 2 x 5 x 2 feet, in ruins.

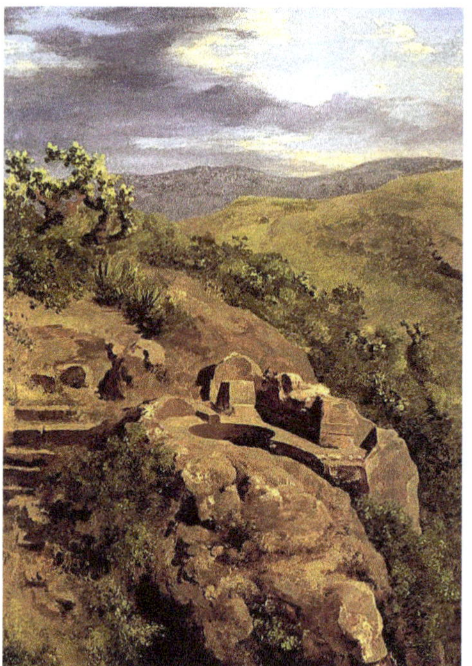

Baño de Nezahualcoyotl, José María Velasco, oil on canvas, 17 x 24 inches, 1878.

But Urrutia adds to the story, for unlike Nezahualcoyotl and Cuauhtémoc, he knows that the eagle descends and then rises in a story of decimation and triumph. A phoenix out of the ashes, the eagle becomes the national symbol of Mexico. Originating from pre-Hispanic times, the first Mexican garden emerges, unique in form and purpose, in a tradition of indigenous creation. Urrutia pays tribute to the roots of the Mexican garden and brings this interesting relic of the past into the future via Miraflores centuries later, leaving it for rediscovery.

For Urrutia and other Mexican intellectuals of the time, Nezahualcoyotl was as revered for his enduring songs and poetry as he was for his engineering, architecture, and success as a ruler. The polyglot king reunited a wide swath of cities in 1431 and ruled until 1472. He engineered kilometers-long waterways to bring the water supply from Chapultepec to Texcoco, where he built his massive palace and gardens.

Although the Spanish destroyed or buried the extensive features of the Texcocoan botanical garden and palace around 1539, folklore surrounding the garden persisted in songs and poetry until it could be unearthed and restored.[102] Urrutia and his friends viewed the artist José María Velasco's painting of a landscape depicting one of the garden's "baths" with its famous curving staircase, acquired by the Museo Nacional in 1878. The large rounded stone pool, designed for ritual purification and perhaps for controlling water flow, appears to be oval because of the artist's perspective. »The shape repeats in Urrutia's swimming pool.«[103] In a letter to Urrutia, Tablada likens the doctor to a son of the Toltec king: "You are a legitimate son of Nezahualcoyotl, who made Texcoco the most admirable indigenous civic center, and the royal poet inherited the poetic gift that makes him create beauty around him and make poems, hymns, not with literary elements, but with earth, water, trees and flowers!"[104]

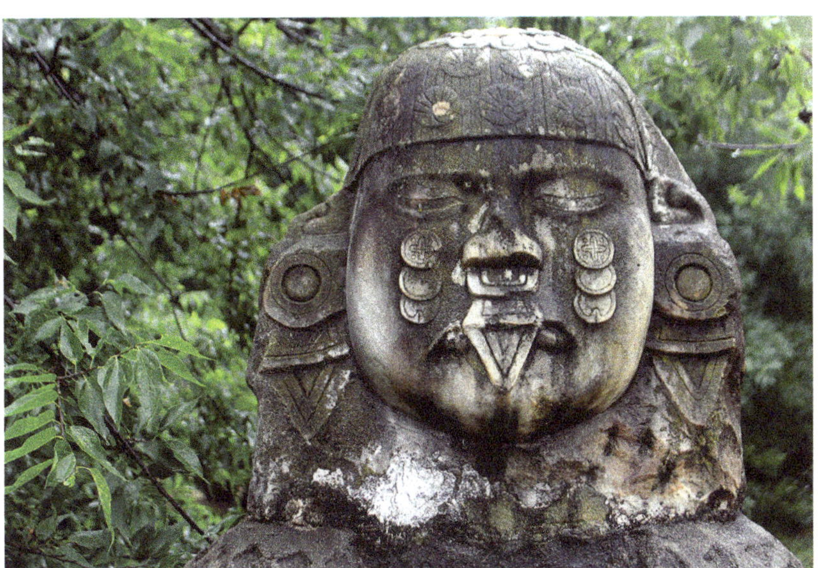

Coyolxauhqui, 20 x 18 x 12 inches.

Coyolxauhqui, the Moon Goddess

A pedestal to the southeast of the *baño* contains a replica of the sculptured head of Coyolxauhqui. The Museo Nacional holds the original, dated between 1450 and 1521.[105] Her Nahuatl name means "face with bells," an easily identifiable feature, and the Miraflores version is signed by artist Sanchez Lopez.[106] Apparently carved from a single piece of stone, the sculpture has distinctive features—both the original and replica include "bells" on each cheek, a featherlike headdress, and ear decorations. Even the back of the head contains exquisite detail.

The original sculpture features elaborate carving on all sides. The museum acquired the sculpture in 1830, just after its discovery near the Templo Mayor in Mexico City. Highly cherished when it was unearthed, it became known as one of the premier Aztec images of Urrutia's day.[107] Originally mystified about her exact identity, archaeologists only correctly identified her as the moon goddess Coyolxauhqui in 1900.

The folk stories surrounding the moon goddess tell that she opposed her brother Huitzilopochtli (hweet-zeel-oh-POHCHT-lee, often associated with the sun) in the establishment of Tenochtitlán as the Aztec homeland. In one variation she opposed her mother,

Coatlicue (k'waht-LI-kweh), who was pregnant with her brother. Huitzilopochtli defeats Coyolxauhqui by decapitating her. He tosses her head to the sky, and it becomes the moon—or, in some versions, the Milky Way.[108]

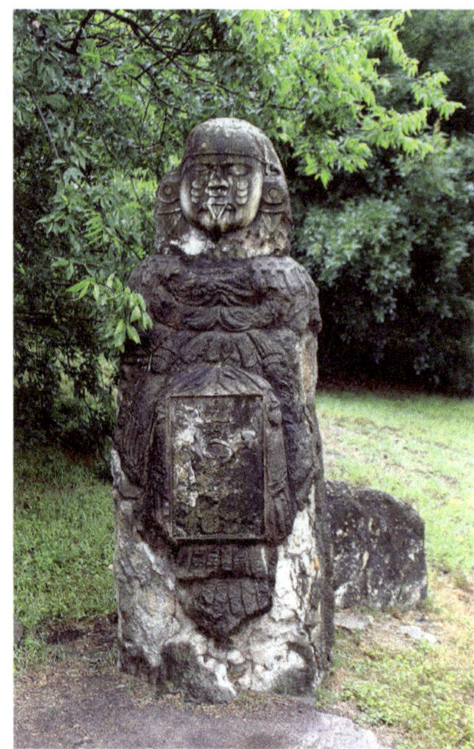

Coyolxauhqui on Coatlicue-inspired base, concrete and stone. Base is 5 x 2 x 2 feet, in ruins.

At Miraflores, ironically, Coyolxauhqui's head sits atop another statue pedestal that crudely appears to represent her mother, Coatlicue.[109] The definitive statue of Coatlicue was discovered in 1790 and identified in 1888. Her folkloric story and symbolism reflect complex opposing forces rooted in life and death. Her terrifying head displays two snake heads facing each other in profile, and yet she gives life to the patron god of the nation.[110] Her duality reflects a message consistent with Urrutia's cultural understanding.

However, rather than the scaly clothing of Coatlicue, the statue upholding Urrutia's Coyolxauhqui appears to have a feathered body, perhaps added later, suggesting a possible conflation of Coatlicue with Quetzalcoatl (ketzahl-COH-ahtl), the feathered serpent—yet a third essential Aztec icon. »The almost completely decayed statue's snakelike form appears to have featured a rectangular plaque, now lost.« The myth of Quetzalcoatl, which references the Aztec duality of flight and earthly dwelling, symbolizes wisdom and knowledge, suggesting that Urrutia's strong value of education originates from his indigenous heritage.

The representation of these major Aztec figures—Cuauhtémoc (1496–1524), Nezahualcoyotl (1402–1472), Coyolxauhqui (representing the founding of Tenochtitlán, capital of the Aztec people, in 1325), and Coatlicue (representing the life force of the Aztec people), along with Quetzalcoatl's message of knowledge—within proximity to each other gives this area of the garden powerful import. They suggest a traveling back in Aztec history, as if in a time machine, from the last ruler of the empire, to one of the ultimate creators of Mexican gardens and poetry, to the origins of Aztec culture.

Interestingly, Urrutia chooses this spiritual place of origins, largely unfamiliar to many visitors, for entertaining them. Most had never experienced such overt expressions of indigenous symbolism. Visitors from central Mexico, however, especially those from Urrutia's milieu, found themselves in a place of connection. Here is a perfect crossroads for Mexican and American culture, highlighting the unique and intriguing aspects of Mexico's origin stories—a theatrical setting for newcomers and for those viewing their culture in a new light.

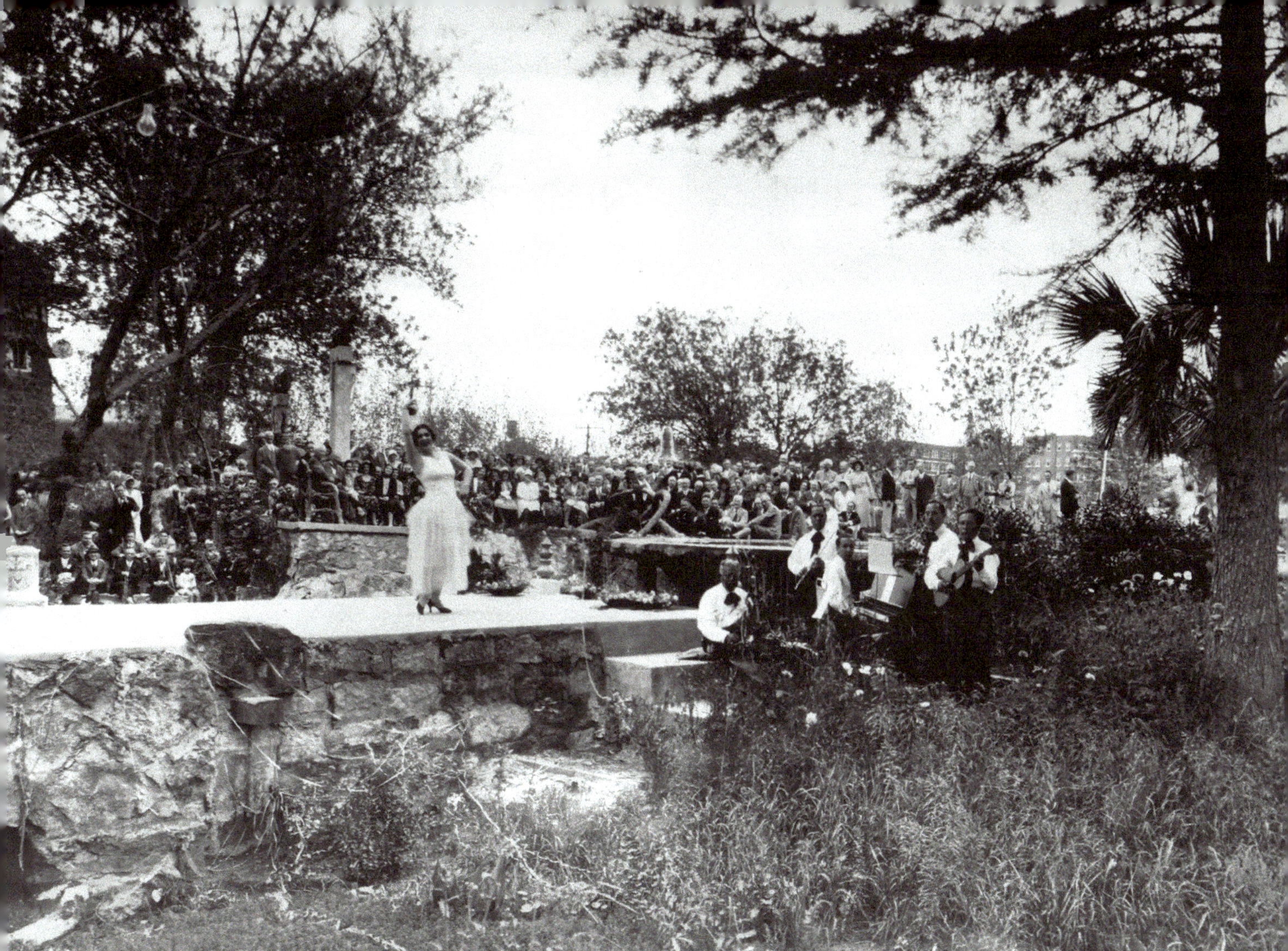

Stage bridge, 43 x 11 feet, now in ruins.

The Pond: Theater for One and for Many

»Beyond the *baño* and esplanade, the stage bridge and pond area create an idyllic and dramatic scene. The long flat bridge stretches from the western edge of the pond toward the *baño* that hovers above it, fronted by water lilies below. Water flows slowly through a small arch under the bridge. Boulders border the edge, tiered steps at each end, and Talavera benches and urns herald the stepped entrance; irises grow along the banks. Hidden behind the bridge but attached to it, a small peninsula juts into the sizable natural pond.«

As a complement to Urrutia's undeniable need for solitude and reflection, the area also serves as a setting for his more public self and provides the perfect environment for entertaining large groups.

By hosting events and gatherings, Urrutia shares his Mexican identity. Strangely, in a society hostile to darker skin, in San Antonio as in Mexico, Urrutia operates as an exception to the rule, and his confidence and stature allow him to move easily between cultures. He seems to feel at home. He hosts his own parties—a grand fiesta for guests or a fundraiser for an organization—and allows people to hold concerts, parties, and other civic and cultural events for various causes in the garden. In the most well-documented photograph of an event in the garden, Cuca Blanco de Sapia-Bosch, the sister-in-law to journalist Federico Allen Hinojosa, performs the Spanish Gypsy flamenco dance Gitenerias at a fiesta for attendees of the American Institute of Architects national conference.

The Stage: A Bridge to Community

For public occasions, the bridge becomes a stage for the entertainment, which might be mariachi music or flamenco dancing by night, or a string quartet by day. The stage bridge's stepped ends become the wings. »The small peninsula behind serves as a backstage area. Guests stand near the pool or along the viewing platform or sit on the faux bois benches. One, a bench seemingly made from cut logs, is on the platform.« Another one, the fallen tree bench, is near the pool. More guests stand along the short, curved, faux bois plank staircase. »Still others sit on a concrete wall of stepped seating to view the stage bridge across the narrow leg of the pond.«

Three Talavera and concrete benches reside on the stage bridge itself. »One of the benches, Banca de la Vida, nestles into the landscape downstage near the entry steps. Circular plates decorate the bench, which extends east to west along the lily pond.« The two other benches, really a two-sided bench, are higher and more visible from the opposite bank. The bench on the stage-side takes a couch shape, with massive arms featuring striking tile combinations of yellow and black fleur-de-lis and vine against a bold interlocking pattern of blue and white. »On the backstage side, facing the pond to the south, a grand collection of fanciful plates set among intricate floral patterns portrays a scene of serpents, a minstrel, and a king.«

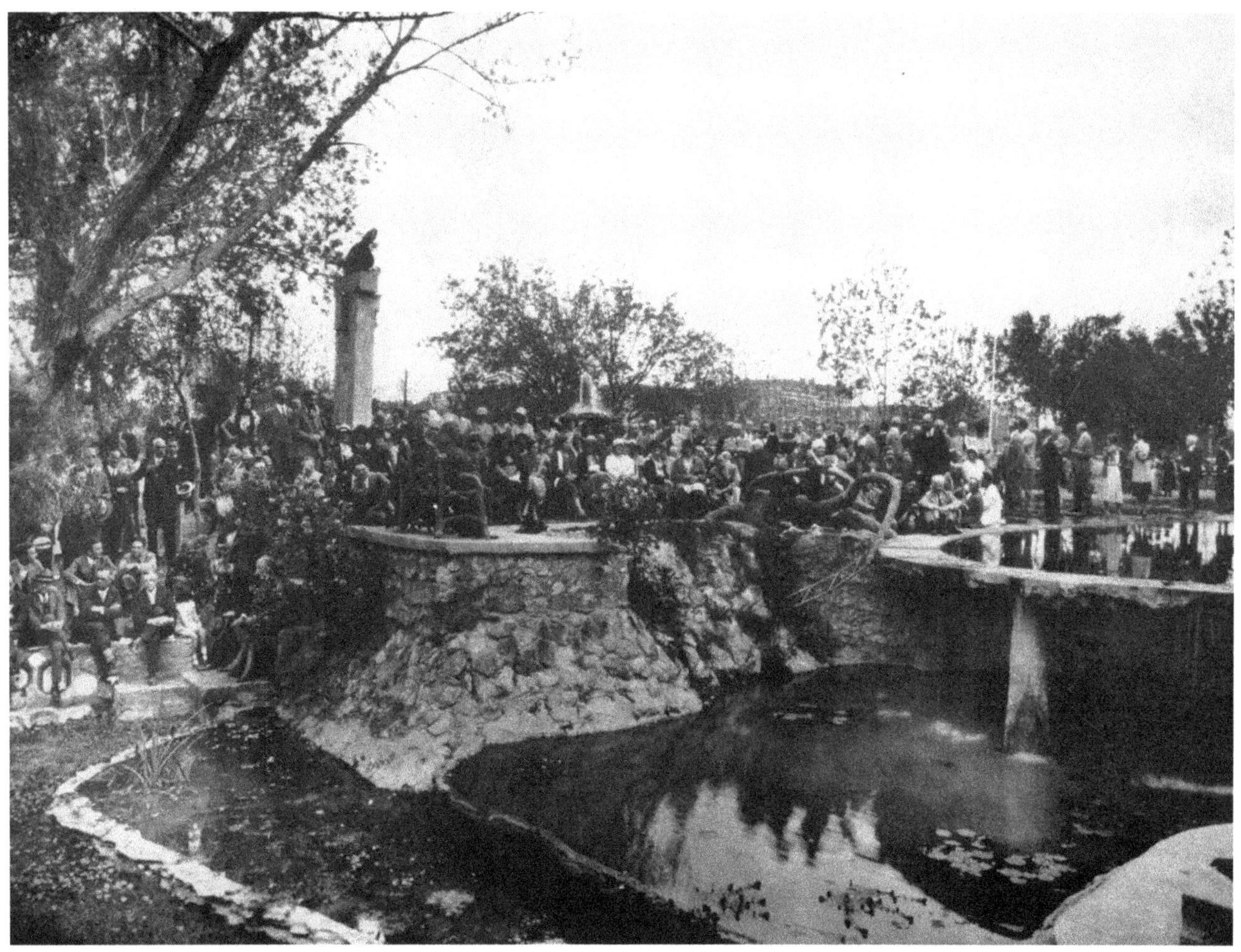

Viewing platform area. From left, concrete stepped seating (nonextant); concrete and stone viewing platform (now in ruins), 12 x 7 feet, with *trabajo rústico* log bench (missing); *trabajo rústico* fallen tree bench; and *baño* and pond (nonextant).

Banca de la Vida, 18 x 42 x 18 inches, right, and two-sided bench, 3 x 6 x 4 feet, below left and right; now in ruins.

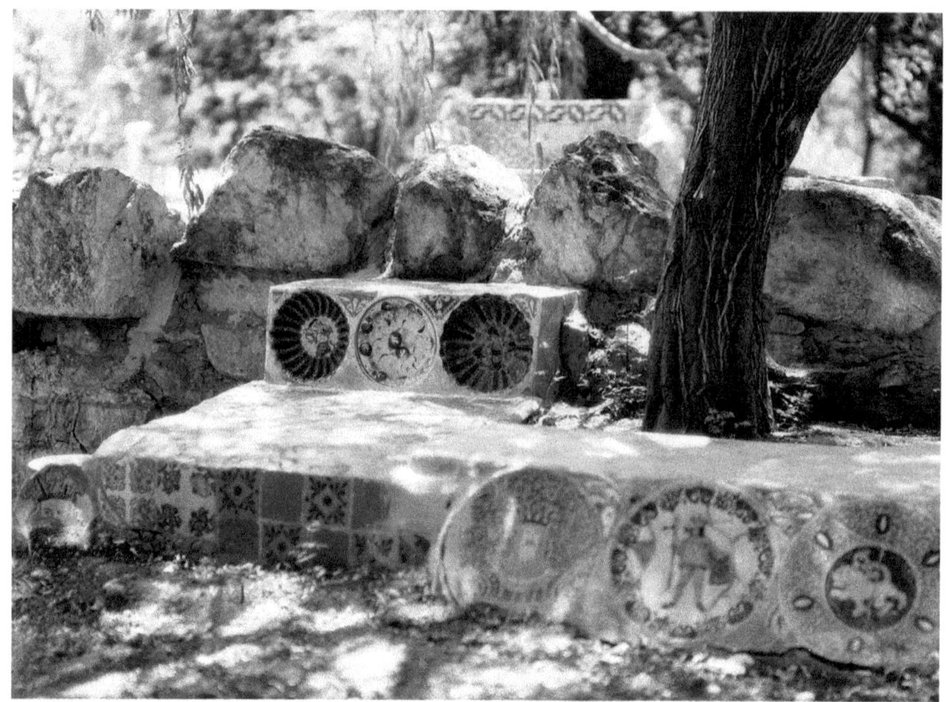

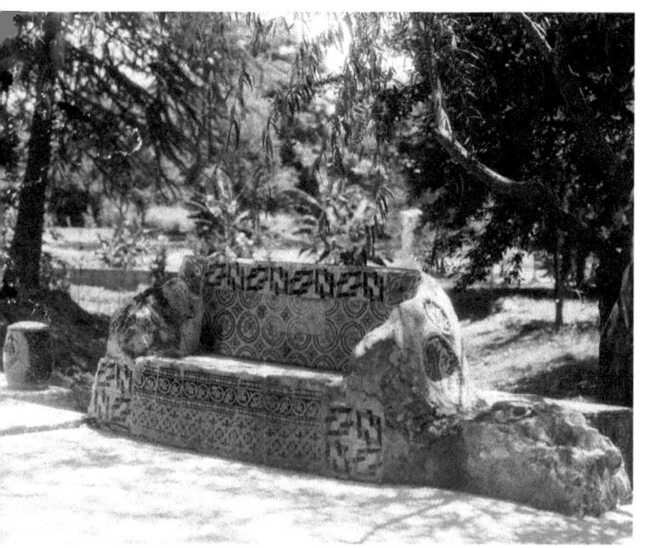

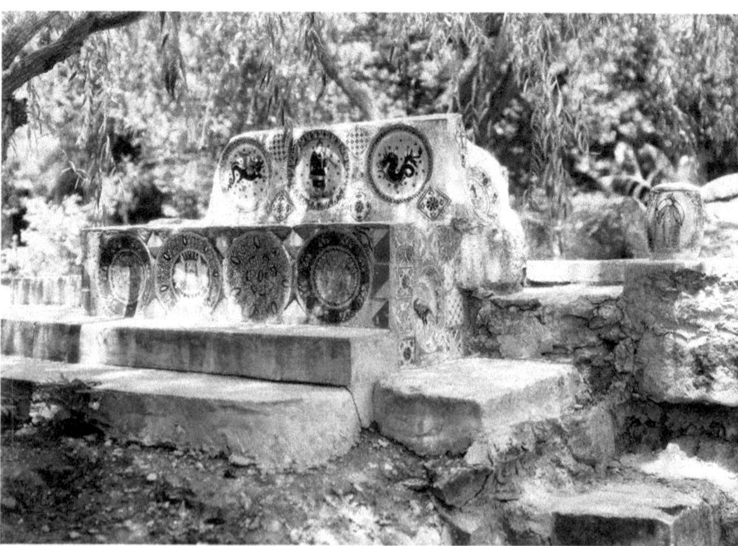

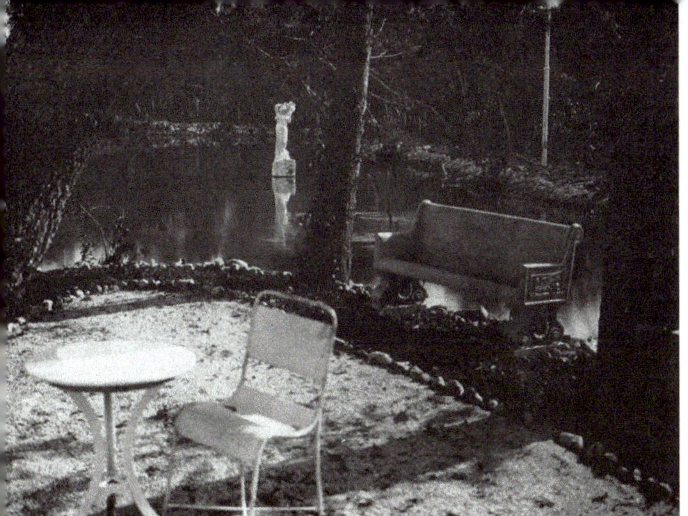

La Isla Luz, nonextant.

La Isla Luz Fernández: Poetry and Waterfalls

In 1921, a month before Urrutia purchased the garden land, Luz, his first wife and mother of twelve children, passed away after a lingering illness. In his eulogy, Urrutia's friend and fellow exile Nemesio García Naranjo, said: "She was the faithful companion, her husband's tireless collaborator, the loving framework supporting his walls when they wavered . . . A comforting prayer in his pain, a cloud of incense on the altar, a cape of glory in the *Te deum!* . . . She was always by his side, like the crystal thread that gives water to the oak, like the pious grove that gives shade to the coffee."[111]

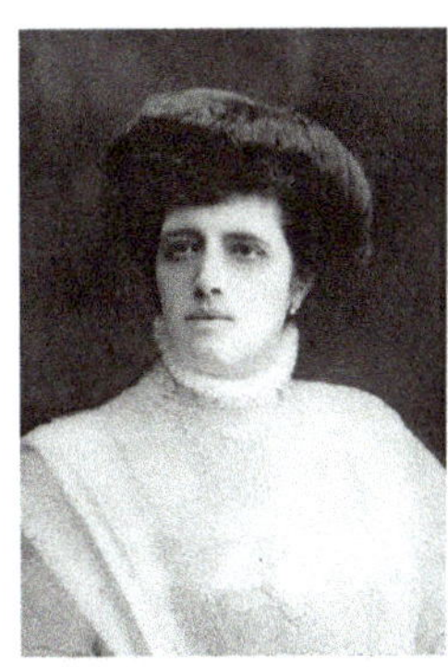

Luz Fernández Urrutia (1878–1921), shown ca. 1905.

»On a quiet day the backstage area serves as a calm, contemplative peninsula. Orderly stones border La Isla Luz, with a solitary bench under a shade tree. This spot is named for Luz. Looking south, a bright white statue of a woman carrying water on her shoulder stands displayed in the pond.« Urrutia identifies the water carrier as crafted by Panduro, and it reminds him of Guadalajara.[112] Perhaps the city is special to the Urrutia family or is a reference to the location where Panduro worked; its significance to Urrutia is unknown.

»The pond beyond La Isla Luz appears natural. With the rains it can grow quite large, filling the areas to the south and west. It ebbs and flows with the weather, sometimes even overflowing into the area behind the esplanade or shrinking toward the stage bridge and the *baño.*[113] Through the willows on the opposite bank, against a backdrop of ahuehuetes, a linear arbor travels across the field. Urrutia writes: "The pond is full of poetry with waterfalls and vegetation brought especially from Texcoco and Xochimilco. The bridge and the island that give access to the pond are surrounded by a beautiful plantation of Ahuehuetes."«[114]

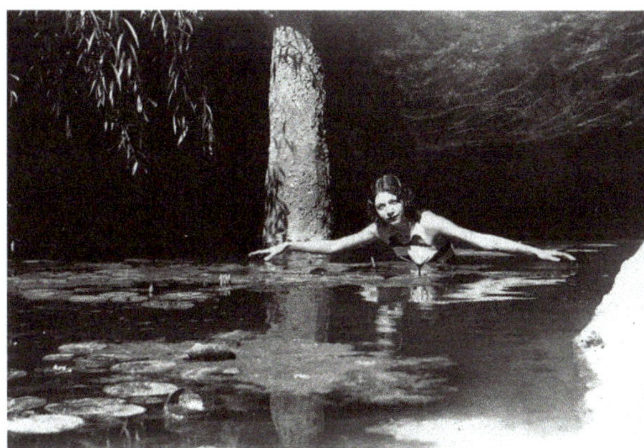

A Urrutia family member, most likely Dolores, one of Urrutia and Luz's daughters, ca. 1925.

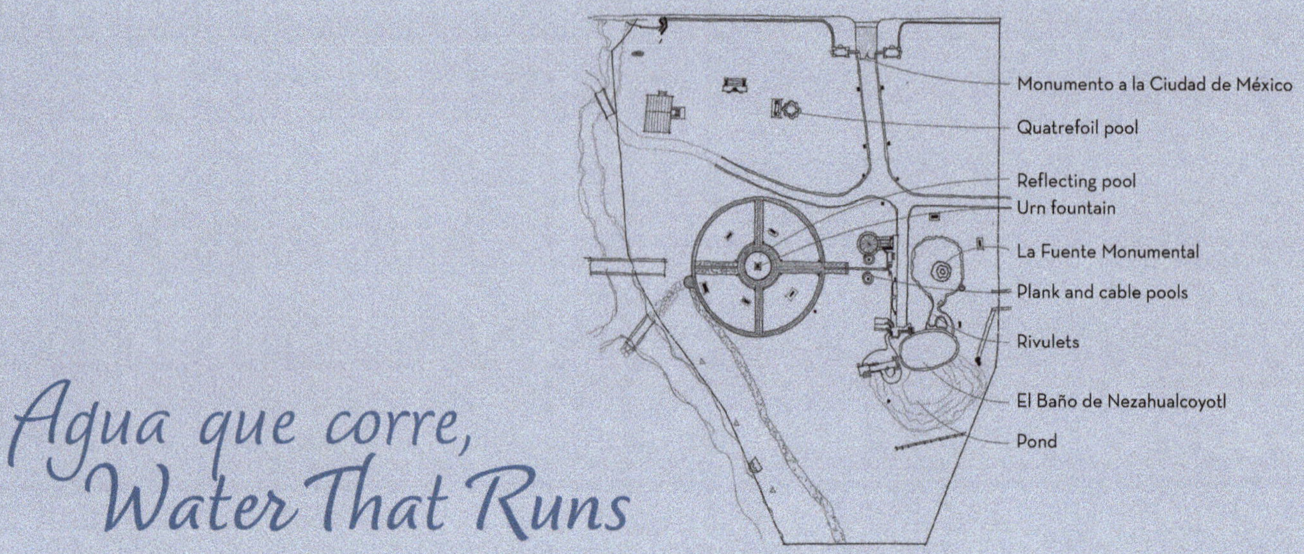

Agua que corre, Water That Runs

The importance of water in Urrutia's life cannot be underestimated. He writes that the name Urrutia, with its Basque origins, signifies *agua que corre*, or water that runs; the Basque word "ur" means water, and "utia" means use or utility.[115] He also conjures tales of a Don Juan Urrutia, whom he characterizes as a colonizer of indigenous tribes along the San Antonio River at the location of Miraflores. According to legend, because of his peaceful ways with indigenous people, Juan Urrutia achieved the rank of marquis in the Mexican state of Querétaro. He is revered for building the city of Querétaro's impressive aqueduct system and is memorialized there in a statue.[116]

Urrutia asserts that a found artifact evidences Don Juan's presence in the garden: "When excavating to build my pool, I found a large stone that says: Year 1716—Here the first mass was held."[117] The granite stone may be a three-hundred-year-old artifact, or the doctor may have commissioned it to commemorate an event significant to him. What was the "first mass" he references? Certainly not San Antonio's first mass, which occurred as early as 1691, and it cannot refer to San Antonio's 1718 establishment as a civilian settlement.

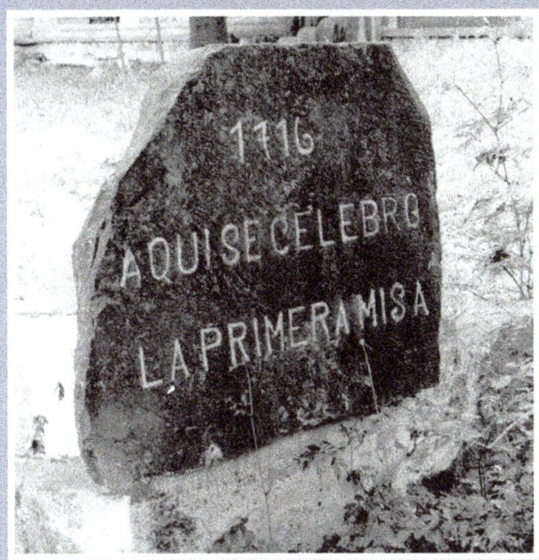

The 1716 stone, 25 x 30 x 5 inches.

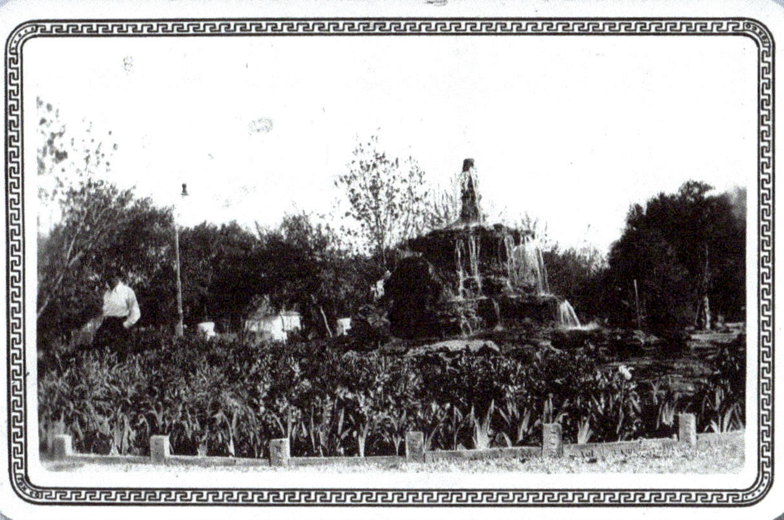

View of Fuente Monumental, Urrutia at left.

Urrutia associates 1716 with the first Franciscan mass in Spanish Texas, which may refer to the founding of San Antonio's Mission San Francisco de la Espada.[118] The mission provided water to surrounding communities via its extensive aqueduct system. Urrutia's childhood among Xochimilco's unique water networks, and the Franciscan church there where he was baptized, would have drawn him to the mission's similar role. The year 1716 was important for Texas's first mission, finally established after twenty years of rejection by indigenous groups, flooding, drought, and disease.

Whether fact or fable, the story of Don Juan Urrutia at Miraflores reflects Urrutia's affinity for water—its importance to the citizens of his hometown, as a symbol of life and nourishment, and as the source of his love for gardening. »For Urrutia, the waterways of Miraflores, fueled by the artesian springs of the San Antonio River, are holy ground.«

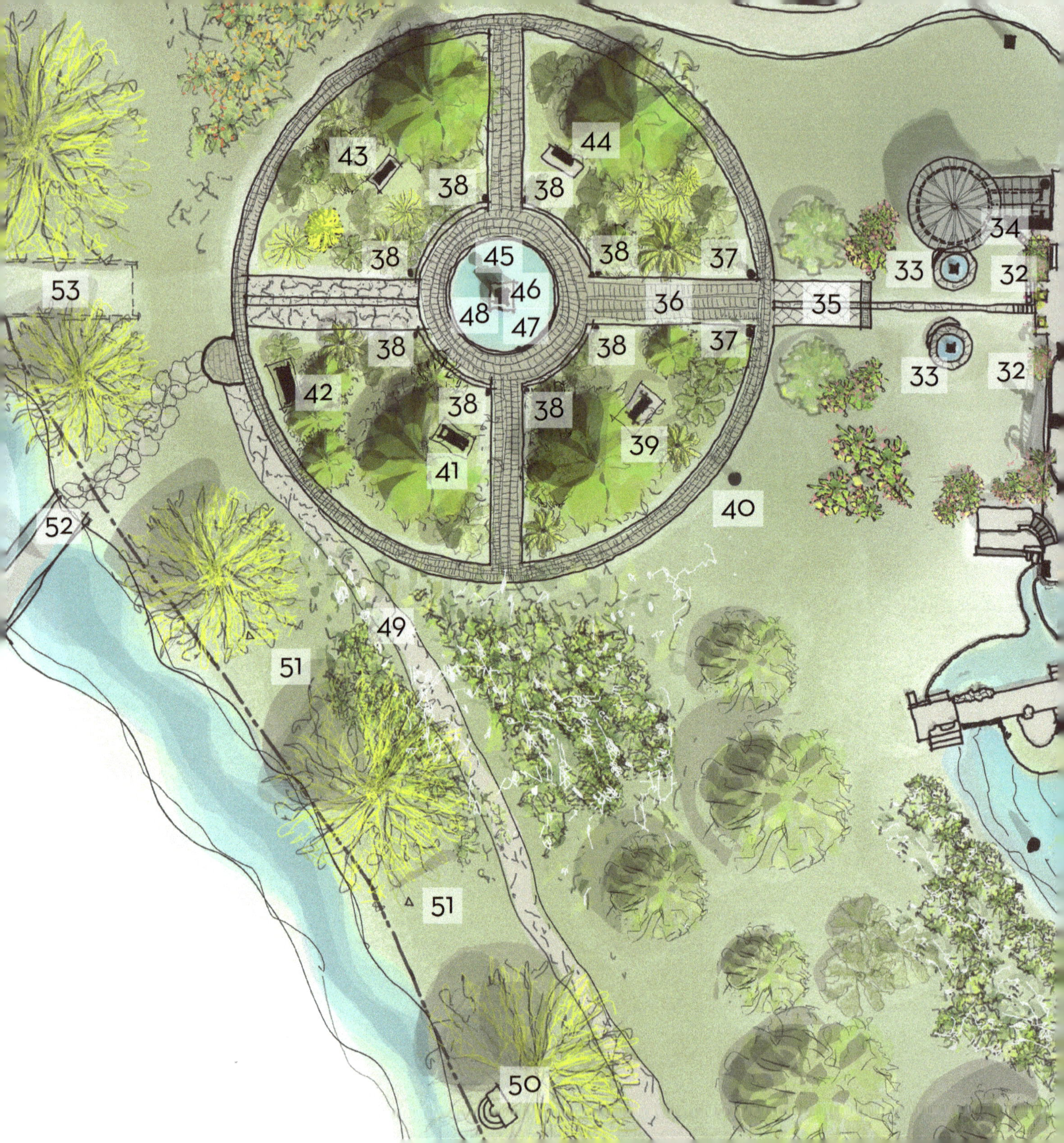

44
43
38
38
34
45
38
38
37
33
32
53
46
36
35
48
47
38
38
37
33
32
42
38
38
39
41
40
52
49
51
51
50

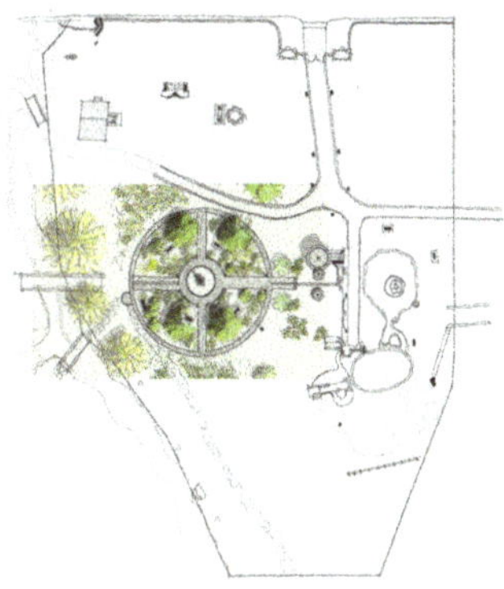

The Compass of Centuries

»Returning along the esplanade, near the tower entry, a small set of stairs leads down to a narrow pathway setting off to the west through a grassy clearing tucked behind the tower. In the 1930s, the path eventually leads to a prairie-like meadow displaying a reflecting pool; in the 1940s, the meadow becomes a garden room—a bricked patio plaza surrounded by high walls of greenery that cloak the shallow circular pool at its center. But first, the rose scent in the verdant antechamber to the reflecting pool area slows the path and beckons time to wait.«

32 Rose courtyard
33 Plank and cable pools
34 Chaise lounge bench
35 Plaza entry
36 Plaza del Centenario
37 Entry urns
38 Intersectional pedestals
39 Inscription bench
40 Diamond urn
41 Double bench
42 Scroll bench
43 Banca del Amor
44 Lost bench
45 Reflecting pool
46 Urn fountain
47 Inscription to Fray Pedro de Gante
48 Dr. Aureliano Urrutia statue
49 Riverside footpath
50 Riverside steps
51 Riverside markers
52 South bridge
53 Park pedestrian bridge (2007)

The Rose Courtyard: A Fragrant Chamber

»In this secluded and shaded chamber, deep in the garden center, a peace awaits. Here, the Virgin of Guadalupe atop the Urrutia arch seems a world away, and yet her presence cannot be ignored. Castilian roses perfume the air as they climb along the esplanade walls. Water trickles from lion's head spigots, performing a quiet lullaby. Two small shallow faux bois plank-and-cable pools catch the streams.[119] Clouds of pink double roses drift across the lawn. They seem to recall the tale of the miraculous roses that fell from Juan Diego's cloak when he revealed the mystery of Our Lady of Guadalupe to the bishop.«

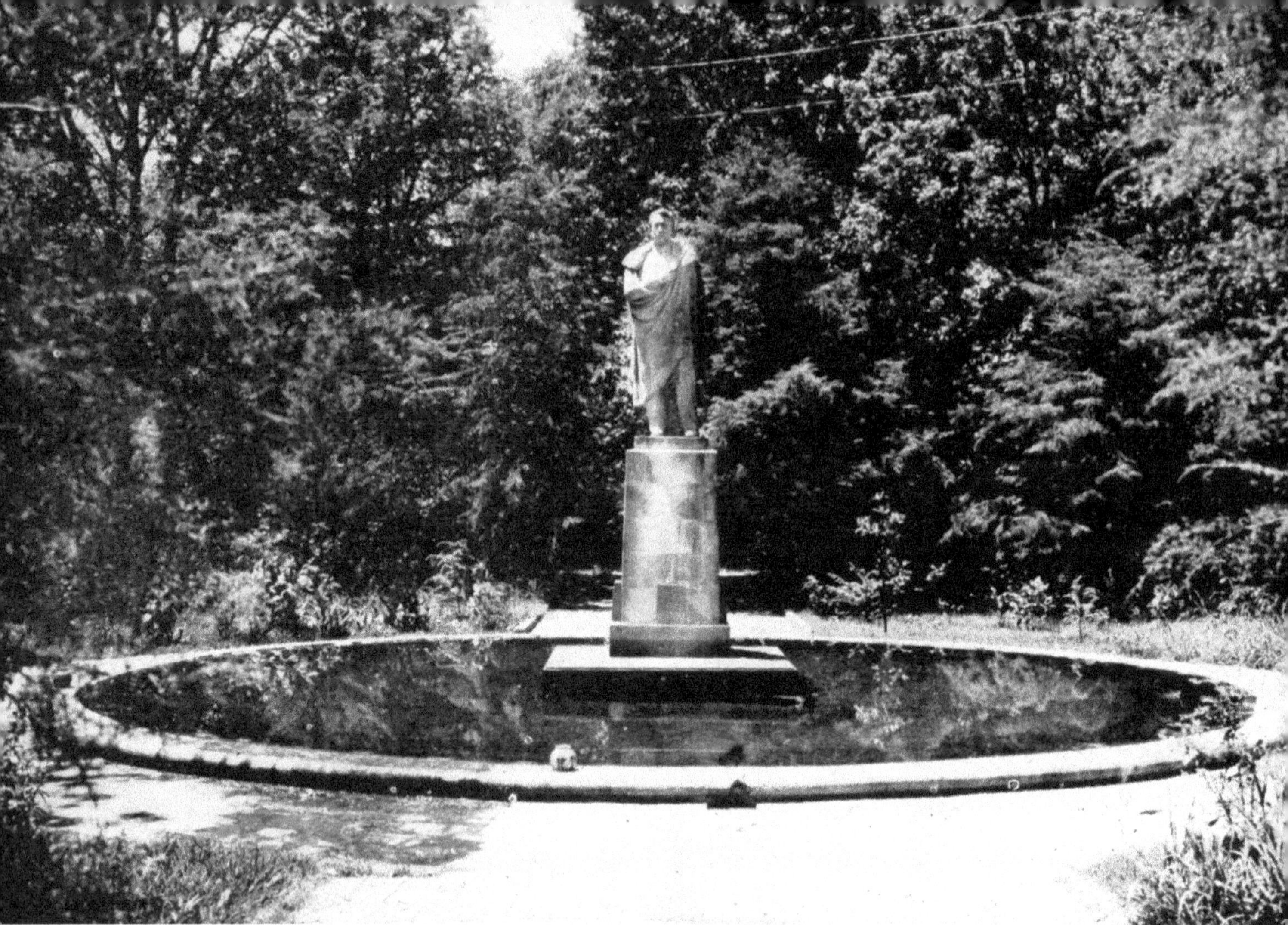

Plaza del Centenario, pool is 37 feet in diameter, plaza as pictured is nonextant, statue restored.

Tucked along the wall of the tower pavilion, a chaise lounge bench evokes a lace-covered divan, perhaps of oriental influence.[120] At its head, a barrel-shaped Talavera vessel turned on its side becomes the pillow, decorated by a regal rampant lion, flanked at the base by two crowned eagles. A table's worth of Talavera plates along its side display a pair of sparring knights, with yet more lions frolicking in a forest of fleurs-de-lis.

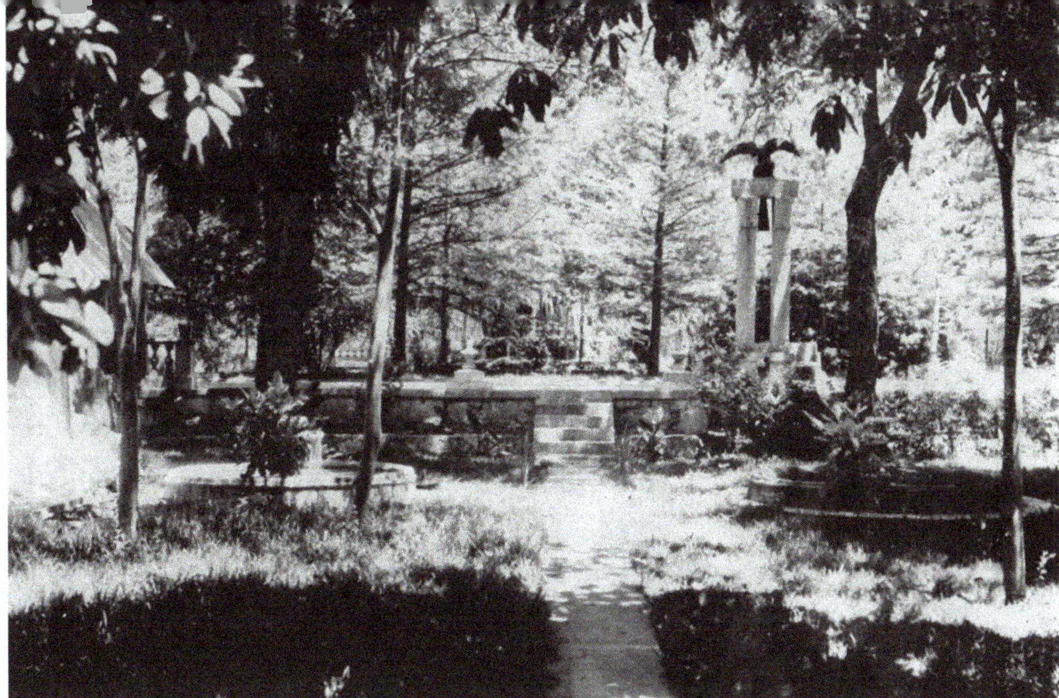

Rose courtyard, 56 x 40 feet, nonextant; with *trabajo rústico* fountain pools, 1 x 9 feet in diameter, now in disrepair.

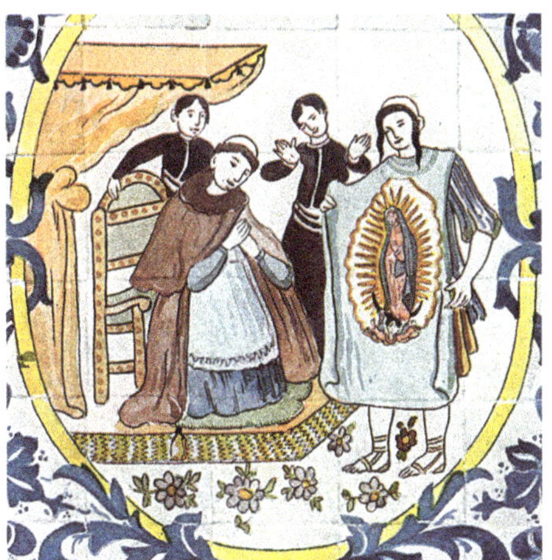

A nineteenth-century Talavera tile panel by Enrique Luis Ventosa in the Church of Guadalupe, Puebla, Mexico, recalls the miracle of the roses.

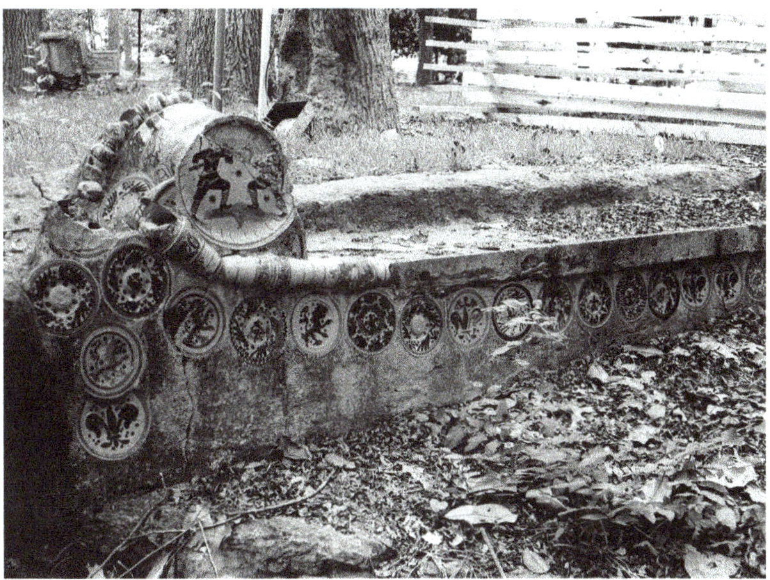

Chaise lounge bench, concrete, Talavera, 3 x 10 x 2 feet, in ruins.

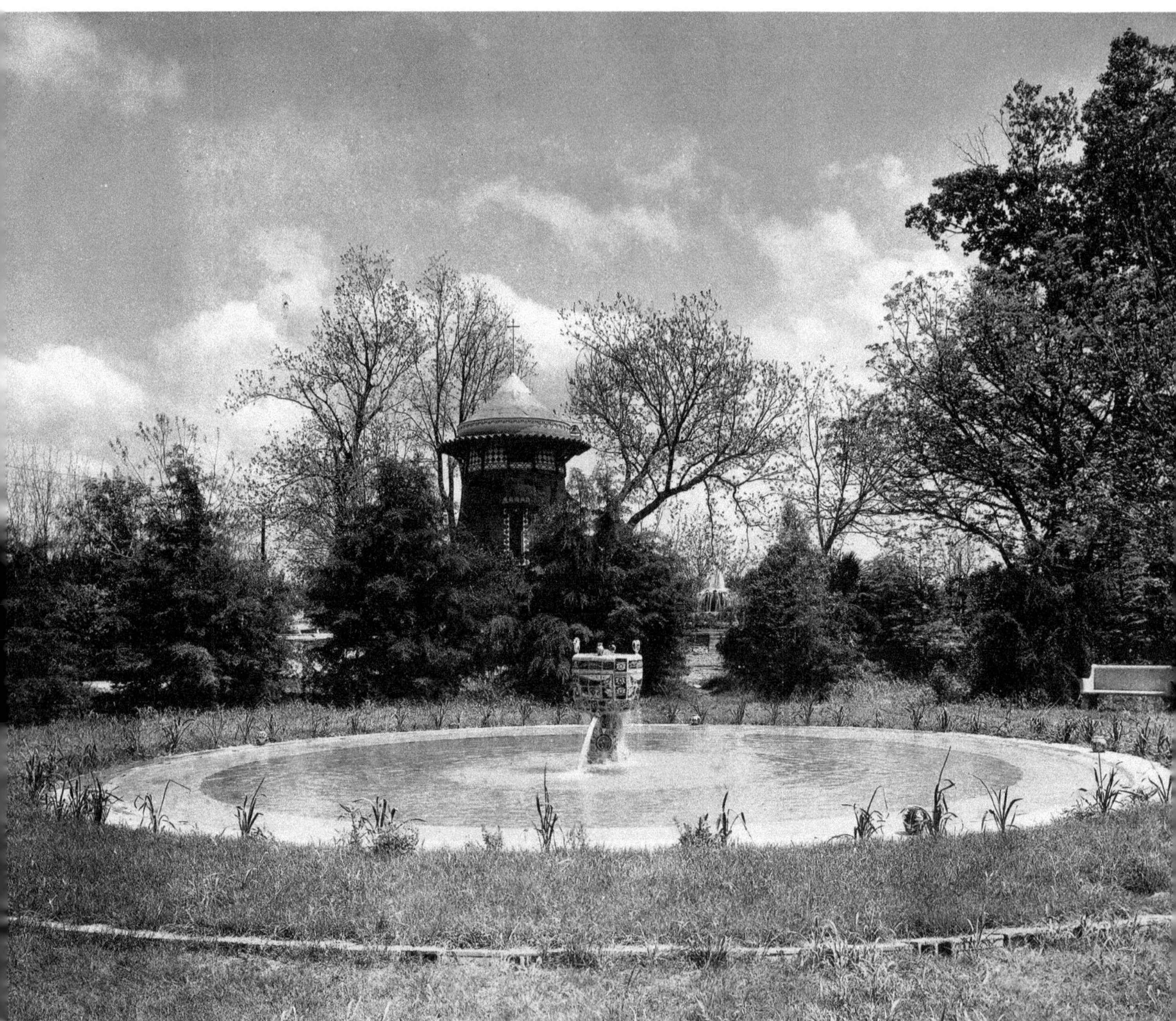

Reflecting pool, with central urn fountain, 4 x 3 feet, nonextant.

The Reflecting Pool: Honoring De Gante

»Beyond the rose courtyard lies a reflecting pool. In the pool's first decade, a large mosaic Talavera urn fountain sometimes sprays a shower from its center, high into the air. It rains back into the still water below.« Urrutia celebrates his fifty-seventh birthday here in 1929 with a large gathering.

The lip of the thirty-foot-wide concrete pool is nearly flush with the ground. A row of *azulejos* echoing the border on the Urrutia arch decorates the inner edge. »Several small oriental-style Talavera jars punctuate the pool's edge.« A single line inscription etched along the foot-wide coping on the pool's southeast quadrant reads:

1530 4th Centenary of the First School of America
Founded by Fr Pedro d'Gante in Texcoco 1930

The inscription recalls Friar Pedro de Gante's founding of the earliest school in North America, memorialized here four centuries later, in 1930, with the installation of the reflecting pool.

Urrutia's use of centennial commemorations, at the Monumento gate and at the Cuauhtémoc statue, and now with De Gante, stresses the complex impact of these dynamic and powerful figures on Mexican history and reality. They also reflect Urrutia's personal experiences, especially the education he received as a young man of indigenous descent. He again presents his understanding and acceptance of the many forces of Mexican culture over time, including Spanish, indigenous, religious, and more. In connecting the reflecting pool's construction with the inscription commemorating De Gante, Urrutia communicates a hope that the garden will serve as a place of education for those who take the time to understand the references, a space to meditate on the flowers of life's complexities and to learn about Mexican history.

As a teacher and philanthropist, Urrutia also dreams that the garden, which sits on a larger plot of undeveloped land, will exist into the future as the grounds adjacent to a medical center yet to be built. He dreams that medical students, nurses, and patients will one day experience the healing forces of these pastoral surroundings as they did at the Sanatorio Urrutia in Coyoacán.

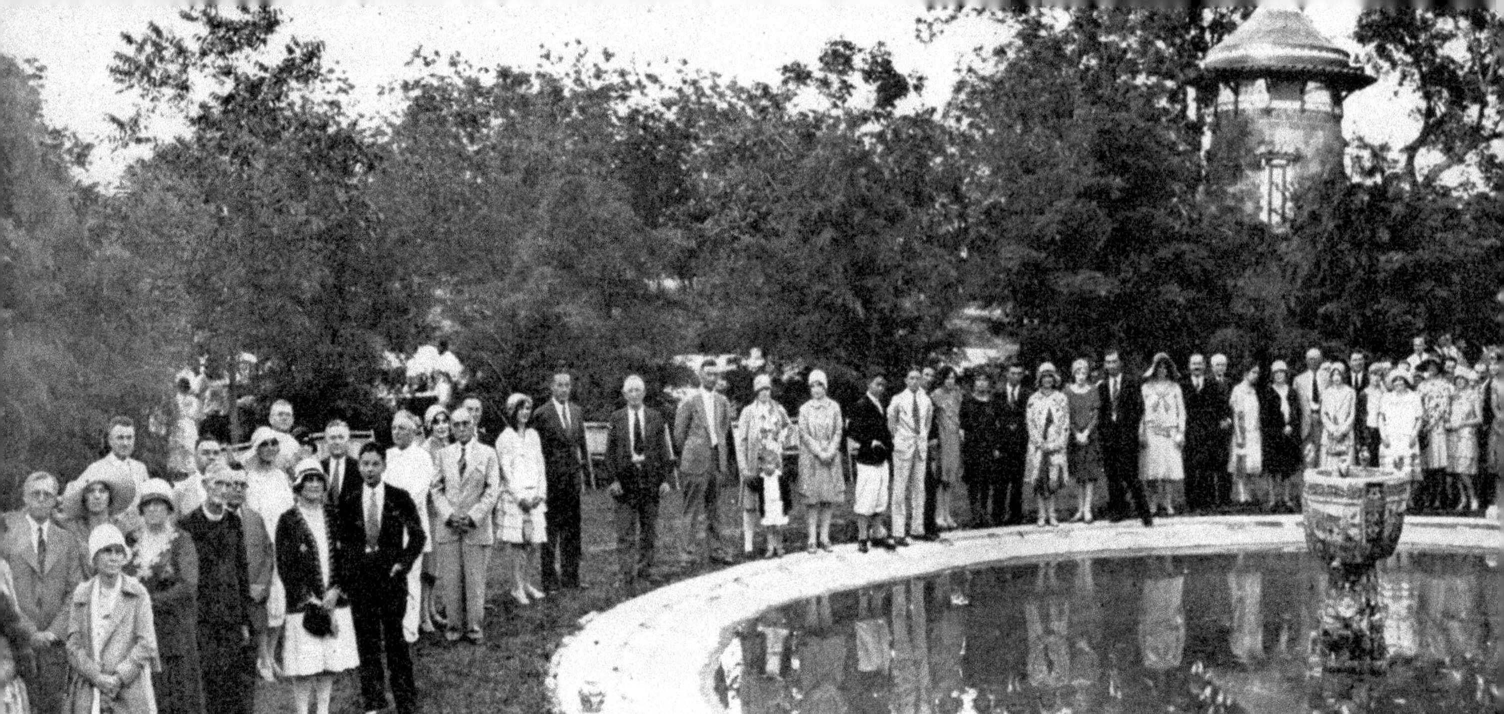

Reflecting pool, at a 1929 gathering.

Plaza del Centenario: A Garden Room

»A decade later, the transformed garden room becomes Plaza del Centenario, referencing De Gante and presciently foretelling Urrutia's own longevity.[121] Patterned brick and inset diamond-shaped concrete stepping stones usher guests to the plaza. Flanked by cypress and magnolia trees, raised cast concrete urns guard the entrance. The urns, decorated with angels and trumpeters, sit atop squared tiled pedestals and are crowned by large taza forms. The space around the reflecting pool has been enclosed with walls of greenery and walkways with a strong directional orientation.« The area may recall De Gante's innovative use of an open-air chapel, built to convert an indigenous population to Catholicism.[122]

Two concentric circular brick walkways, wide enough for two walking side by side, encircle the pool. The inner walkway hugs the lip of the pool's edge. Separated by a deep bed containing thick plantings of trees and shrubs, the outer walkway, nearly

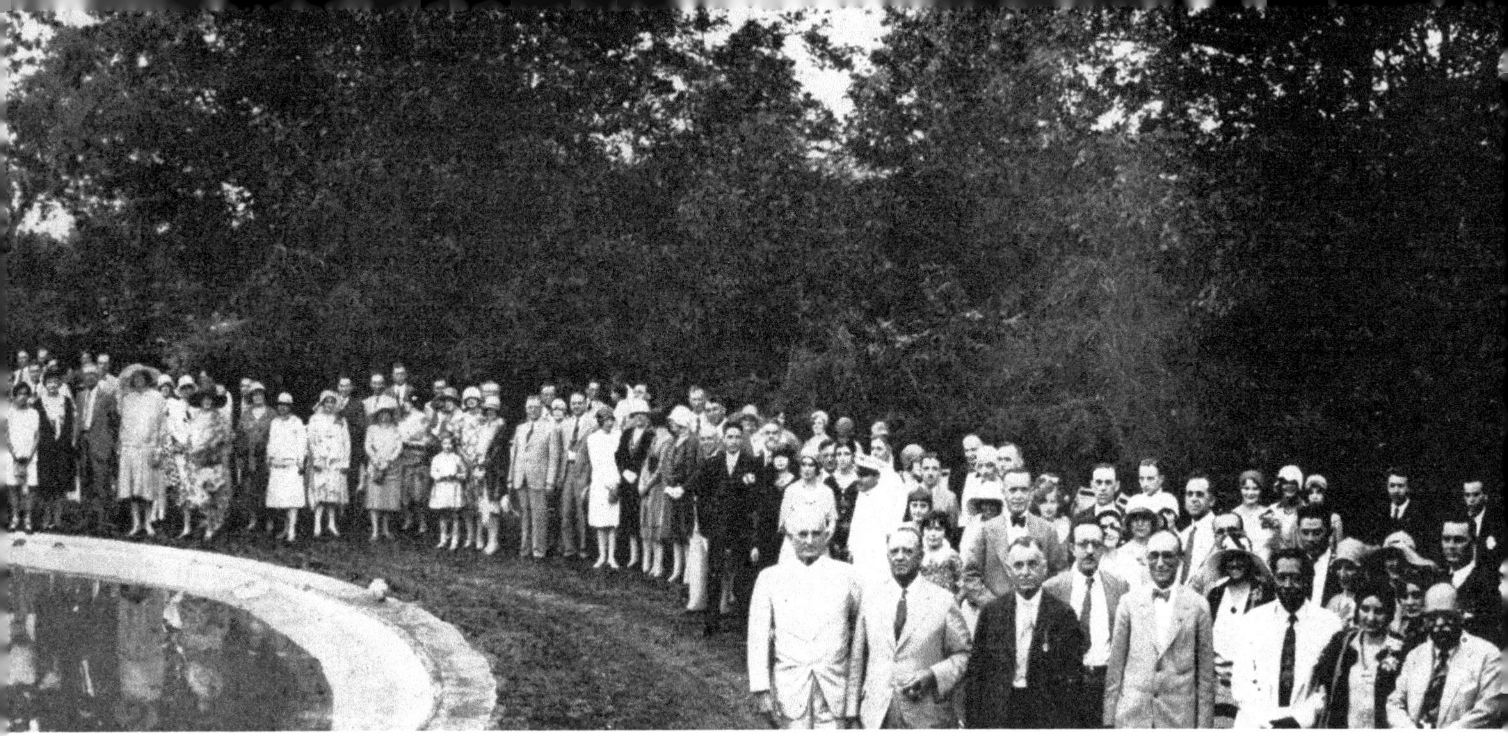

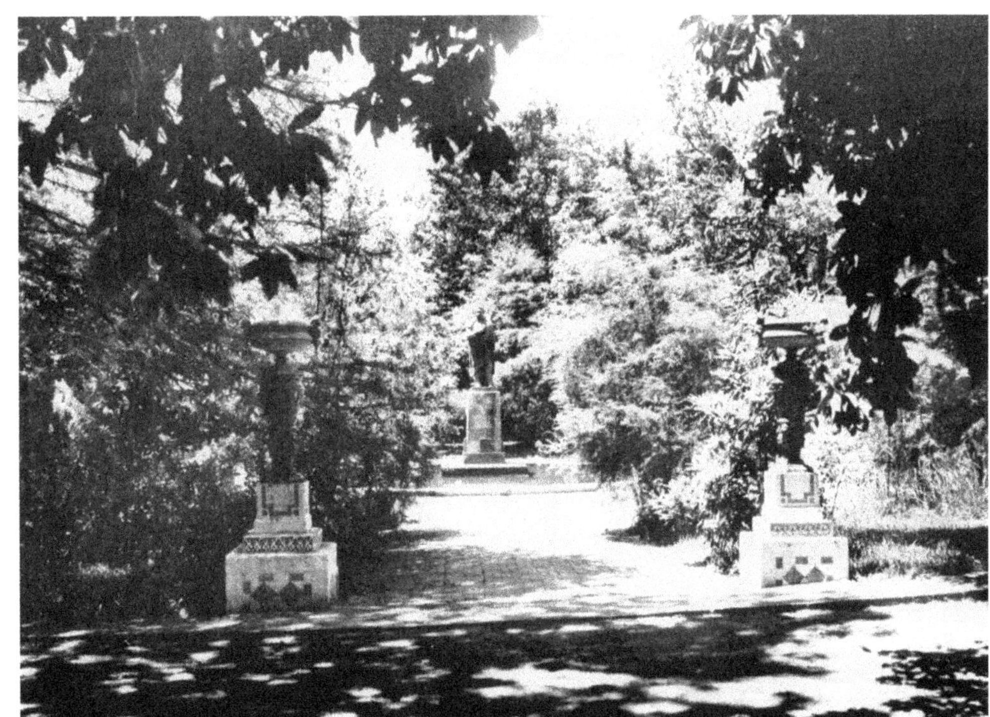

Plaza entrance, nonextant, with pedestals and urns, 6 x 3 x 3 feet; pedestals and urns now in ruins.

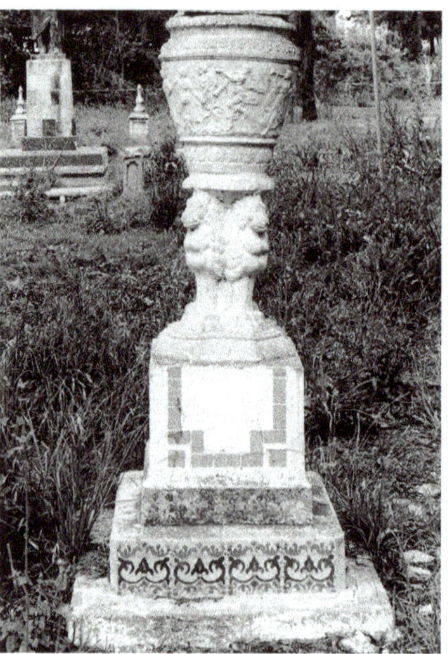

Plaza entry urn, detail.

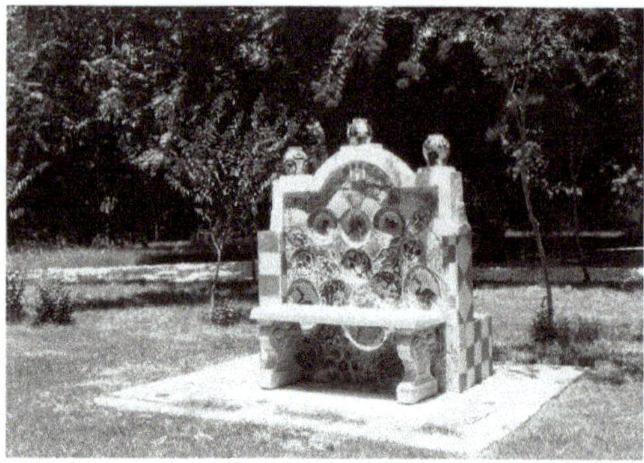

Banca del Amor, 5 x 5 x 3 feet, now in ruins.

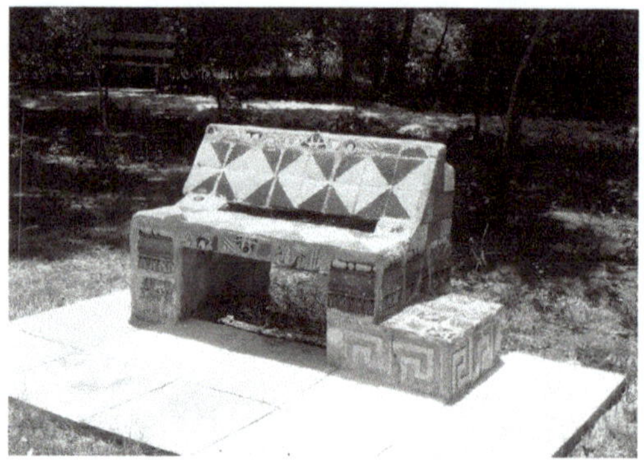

Double bench, concrete and tile, 3 x 4 x 2 feet, nonextant.

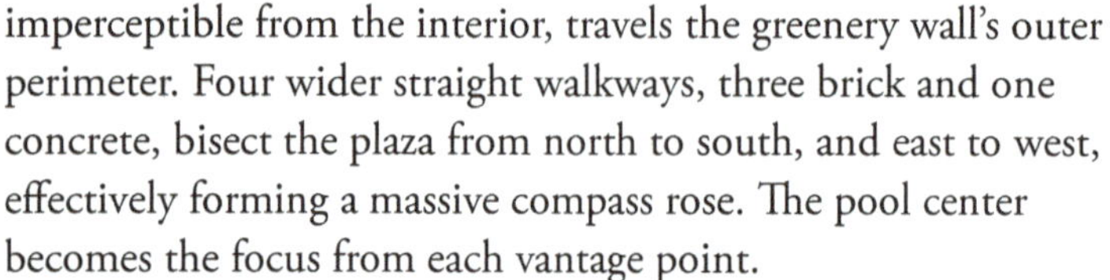

imperceptible from the interior, travels the greenery wall's outer perimeter. Four wider straight walkways, three brick and one concrete, bisect the plaza from north to south, and east to west, effectively forming a massive compass rose. The pool center becomes the focus from each vantage point.

From a bird's-eye view, the plaza, with its entry patio and long sidewalk from the rose courtyard, appears cross-like. Its circular head, the plaza, is in the west toward the river, and its foot, the plaza entry, is in the east mimicking and expanding on the rhythm of the nearby Torre del Califa structure, with its round tower and long pavilion. The four deep garden beds, defined quadrants bounded by the plaza walkways, surround the interior circular walkway, forming the walls of the secluded plaza room. In each bed, benches provide nestled places for dwelling and contemplation.

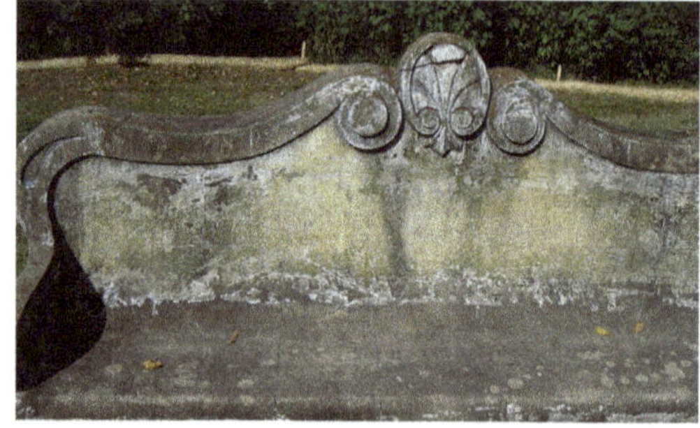

Scroll bench, 4 x 8 x 3 feet, in ruins.

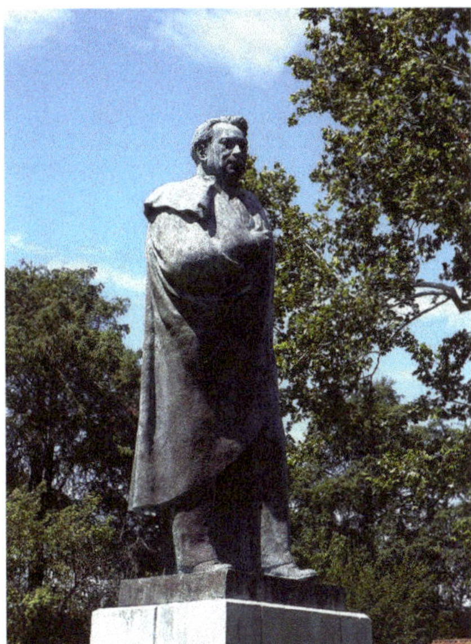

Dr. Aureliano Urrutia, 7 x 3 x 3 feet, shown before restoration.

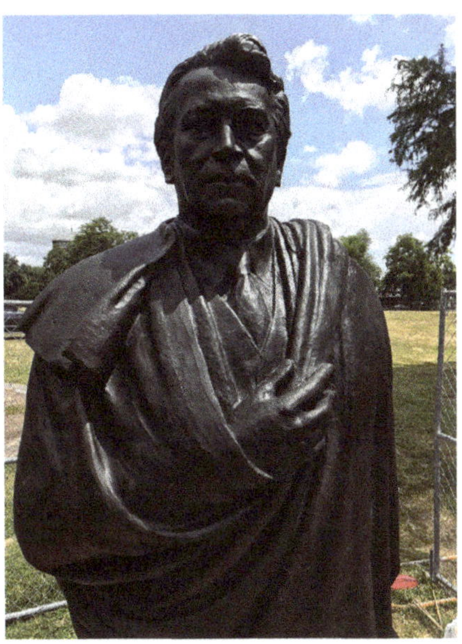

Urrutia statue, detail, restored.

The northwestern quadrant contains the Banca del Amor. Inspired by the tale of a "Chinese Princess who married a Spanish Governor of the Philippines," the seat displays "two sides, one for the weeping Princess, the other for her singing lover." The two, it seems "migrated to Mexico, bringing a potter who taught the Mexicans the color and design of Chinese tile," which is reflected in the Talavera of this bench.[123]

»The southwestern quadrant contains a double bench "for the Calif and his wife," inspired by a bench found in Mexico City's Convent of Merced, once the studio space of Urrutia's friend Dr. Atl.«[124] In the same quadrant, a concrete bench of Mexican design (the scroll bench) displays a beautiful scroll ornament at the center back. »Another concrete seat (the inscription bench) rests in the southeast quadrant facing the pool's inscription, and a fifth bench of unknown design (the lost bench) sits in the northeast quadrant.

More pedestals capped with spires or oriental-style jars mark the inner circular walkway's intersections with the directional passages.« Here, Urrutia often meets garden visitors, especially favorite journalists and writers.

The Urrutia Statue: A Precious Gift

A special gift inspires the creation of the plaza.[125] In 1940, as Urrutia celebrates his forty-fifth year as a physician, Indalecio Sanchez Gavito of Mexico, an attorney and former patient, commissions Mexican sculptor Ignacio Asúnsolo to create a larger-than-life bronze statue of Urrutia. The accomplished surgeon had once operated on Sanchez, removing a bladder tumor considered inoperable by other surgeons.[126]

Although the reflecting pool remains essentially the same, Urrutia replaces the urn fountain at the center with the statue, raised on a pink granite pedestal and facing east toward the garden's central esplanade.[127]

In the statue, Urrutia wears his signature cape. His right hand rests over his heart as he looks over the garden. Photographic portraits inspire or mimic the sculptor's masterful design, but in a flourish all his own Asúnsolo crafts an homage near the sculpture's base, at the foot of Urrutia's cape. Here he inscribes: "This statue represents the wise doctor and eminent surgeon Don Aureliano Urrutia and was made in Mexico in the year 1940 by the sculptor Ignacio Asúnsolo."[128]

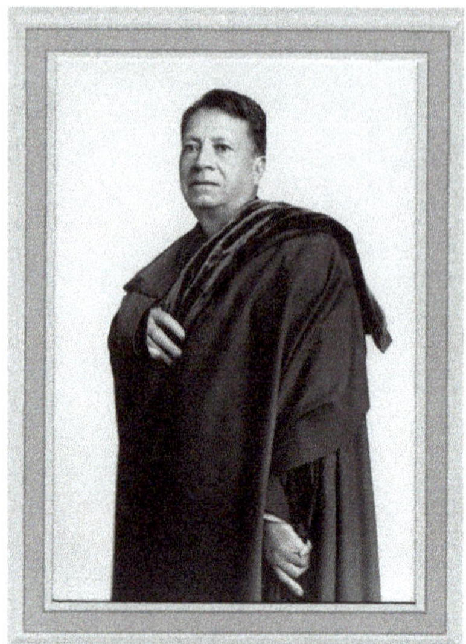

Portrait of Urrutia, 1930s.

In 1926, Urrutia expanded his medical practice by building a pharmacy, clinic, and surgical facilities at Houston and Laredo Streets downtown, near San Pedro Creek. Three sons attended Tulane University's medical school and joined Urrutia in his practice.[129] Daughters owned and ran Farmacia Urrutia; the eldest daughter, Refugio, was the head pharmacist.[130]

During his nearly fifty years as a surgeon in the United States, Urrutia continued his trend of medical advances begun in Mexico. In 1917 he became the first surgeon to have any success separating conjoined twins.[131] He created special instruments, as he often did, for the surgery. The twins were conjoined at the liver. Although ultimately only one twin survived, the landmark surgery drew the attention of the worldwide medical community.[132] In Mexico, Urrutia accomplished pioneering work in surgeries related to the spine and brain as early as 1900. He attributed this progress to his exploration of human anatomy through the study of cadavers, initially controversial but later understood as a pioneering practice essential to the

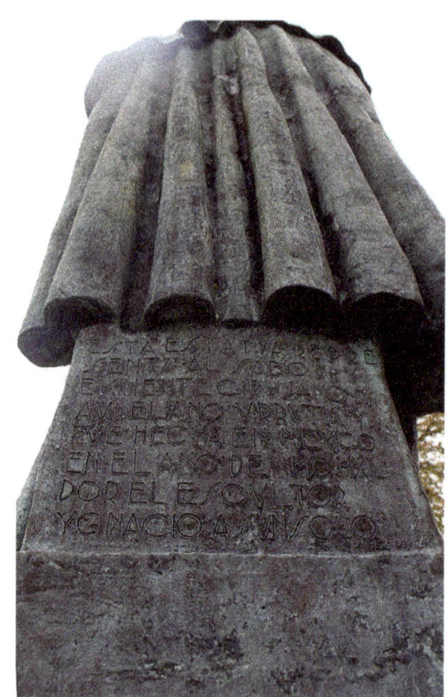

Urrutia statue, detail.

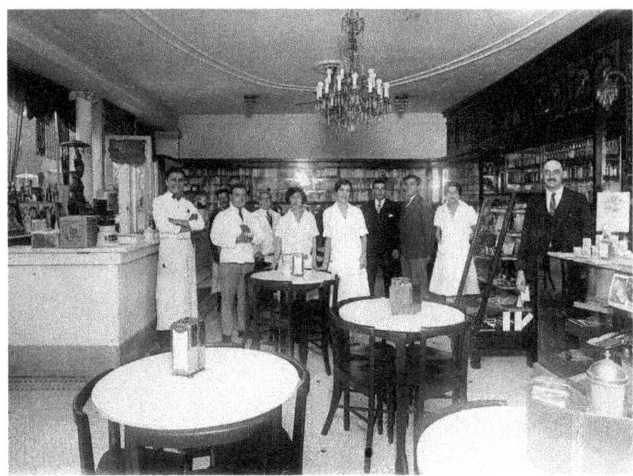

Farmacia Urrutia staff, including three daughters and two sons (at center), Refugio, Alicia, Carlos, Adolfo, and Maria Luisa, ca. 1929.

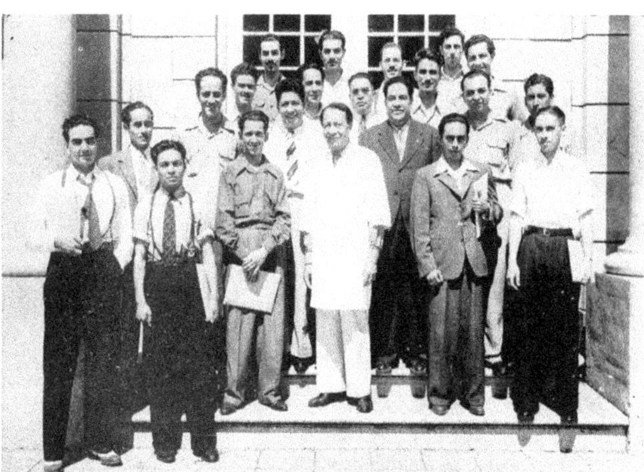

Urrutia with students from Mexico City and Puebla, ca. 1945.

field of medicine.[133] He also instituted the use of cocaine as an anesthetic in Mexico and was an early advocate of fresh air and compassionate treatment for tubercular patients.[134]

Together Urrutia and his progeny served thousands of San Antonians from all walks of life, as well as foreigners who traveled long distances to see him. The practice was especially known for serving patients regardless of their ability to pay. Urrutia also taught and lectured throughout his career on the discipline of surgery in his San Antonio offices, at Santa Rosa Hospital, at the invitation of doctors in other cities, and at international conferences.

Urrutia participated further in the community by opening his home and garden to groups for benefits, concerts, and art tours and often hosted events that bridged the city's Anglo and Mexican communities. He was particularly active in raising awareness and promoting the needs of San Antonio's Mexican immigrant and Mexican American communities, including serving on committees focusing on employment, medical care, and immigration.

A Way along the River: Respecting the Waters

»The westernmost pathway leads from the plaza to the San Antonio River but doesn't quite touch it as the river traverses the property's edge.« Ahuehuete trees planted by

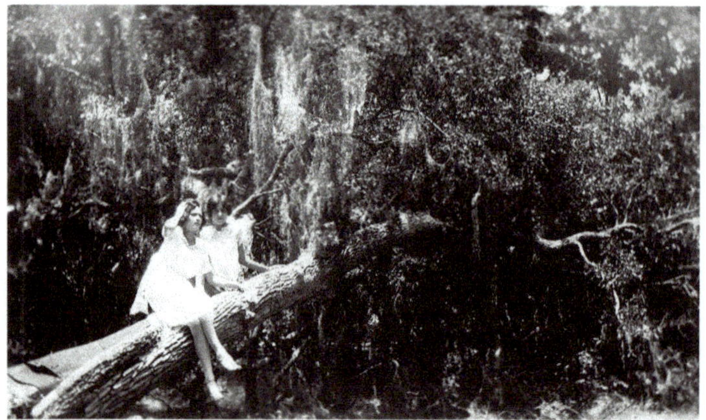

Urrutia daughters, probably Alicia and Emma, at Miraflores, ca. 1921.

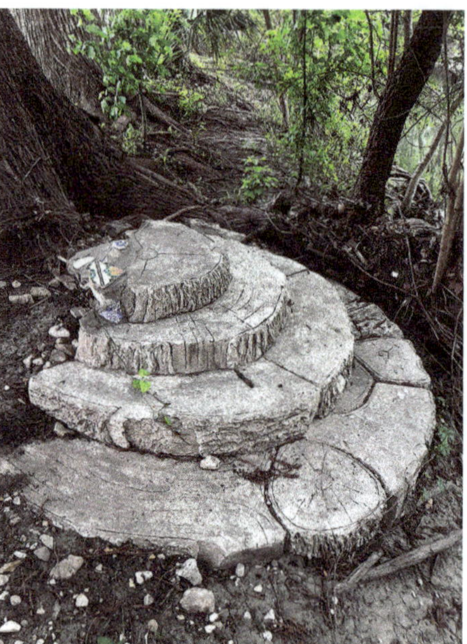

Riverside steps, *trabajo rústico*, 2 x 7 x 6 feet, in ruins.

Urrutia punctuate the bank as the quiet river flows north to south and meanders southeast. »A small semicircular brick transition leads from the outer circle forking into two stone riverside footpaths. The first, a walk toward the river's edge, reveals a spectacular faux bois trestle bridge on the other side paralleling the bank as it crosses over an inlet feeding in from the west.[135] A south bridge continues with little fanfare across the river off to the southwest.[136] The second path turns south, running along and just above the riverbank.« Along the path, triangular faux stone riverside markers emerge from the earth, perhaps lantern shields to light the edge at night. Faux bois steps, each a semicircular platform of crosscut wood sections, lead to the water's edge. »Remnants of Talavera tiles line the top step. Perhaps a traveler's altar or a small seatwall once stood here.«

Along the river, Miraflores is minimally touched by humans but greatly touched by nature. The banks are much as they have always been, unmarred by recent development and the activity of indigenous people, explorers, and early San Antonians centuries earlier. The river's proximity to the public park on the opposite bank, and its source upstream, keep it nestled in safety. Urrutia leaves it be, except for a viewing place and crossings to the other side. The river speaks to him of its own reverent meaning and history. He does not need to interfere. He hopes this might someday be public land that will remain protected.

The Statuary of Miraflores

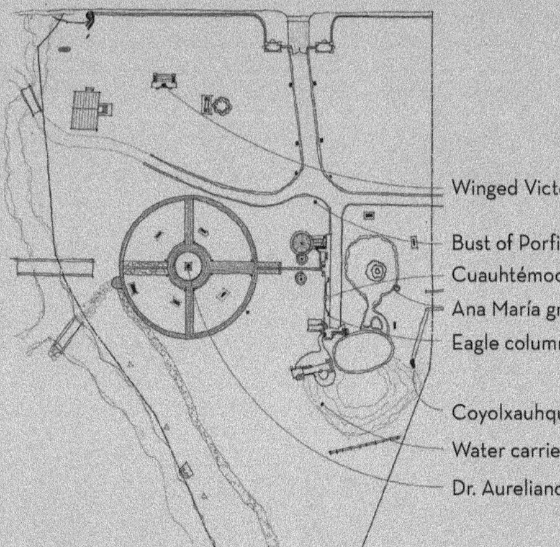

The statues of Miraflores reflect some of the garden's most traditionally European influences and evidence Urrutia's avocation of collecting art. Ironically, European financial control of prerevolutionary Mexico became a cause for conflict, yet European influences on art and architecture survived and emerged as integral parts of the beauty of Mexico today. Incorporating statuary into the garden landscape allowed Urrutia to emphasize people and places that were important to him. The statues in Miraflores have included representations of:

- **»Porfirio Díaz.«** Symbolizing the positive values of prerevolutionary Mexico. Attributed to Pantaleón Panduro, early twentieth century.
- **Ana María grotto.** A tribute to a daughter, Ana María, named on a plaque, with two cherubs in the grotto; this was once the center of a fountain. Attributed to Dionicio Rodríguez, 1925.
- **Cuauhtémoc.** Symbolizing indigenous leadership. L. L. Sanchez, 1921.
- **»Eagle columns.«** Symbolizing the founding of Tenochtitlán. L. L. Sanchez, 1921.
- **Coyolxauhqui.** Symbolizing indigenous origins. Sanchez Lopez, aka L. L. Sanchez, ca. 1920.
- **»Water carrier.«** Symbolizing the beauty of Guadalajara. Attributed to Pantaleón Panduro, early twentieth century.
- **Urrutia.** A tribute to Urrutia's medical career, as commemorated by a grateful patient. Ignacio Asúnsolo, ca. 1940.
- **Mother and Child.** Relief, depicting the Madonna and Christ child, date unknown.
- **Samothrace.** Depicting the European value of appreciating the beauty of the ancient world, and evoking the Winged Victory statue on Reforma Boulevard in Mexico City, ca. 1920.

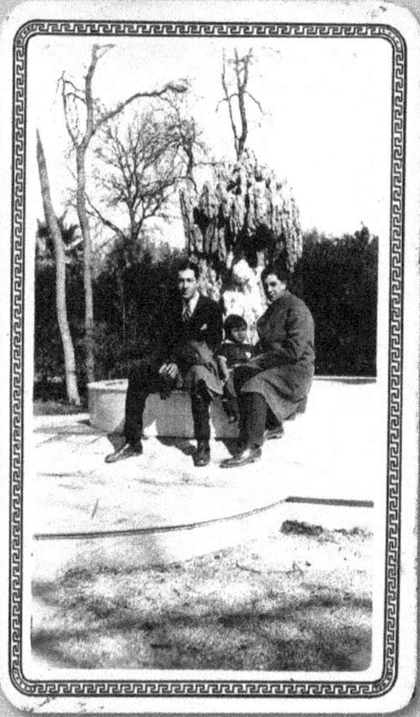

Ana María grotto fountain, before being moved from Quinta Urrutia to Miraflores. Adolfo Urrutia, right, and Carlos Urrutia, left, ca. 1925.

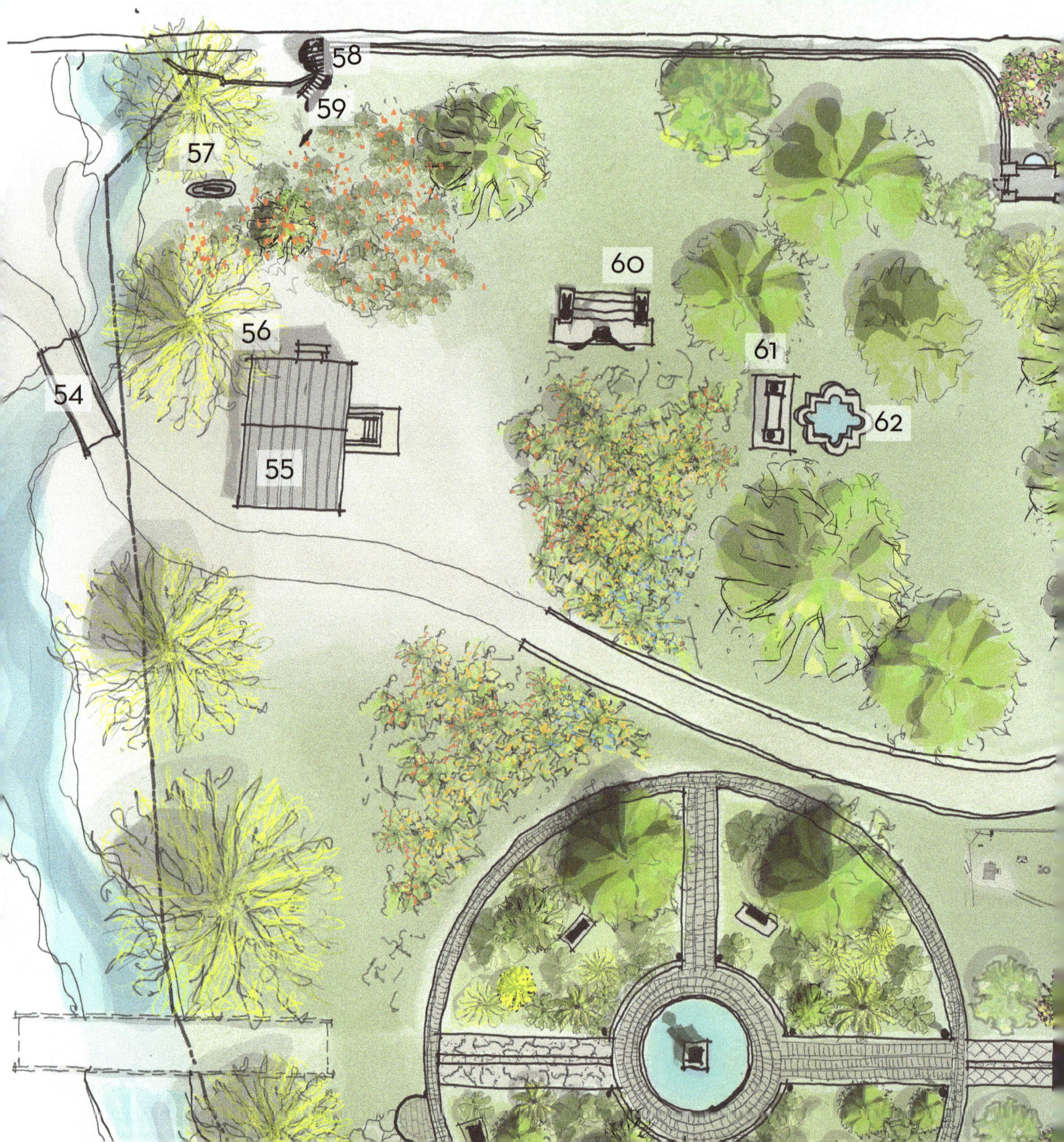
58
59
57
60
56
54
61
62
55

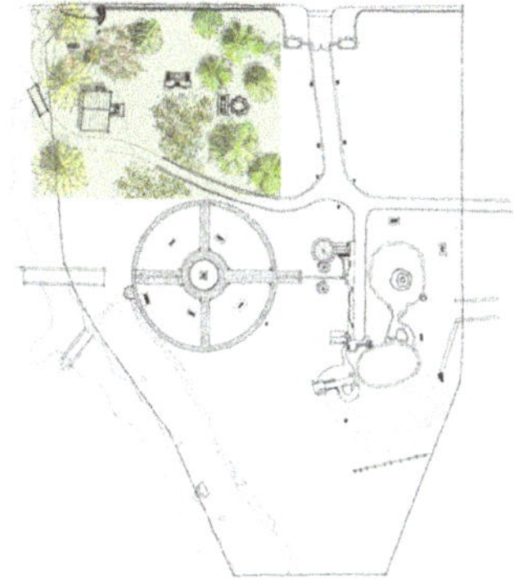

La Quinta Maria

Along the river as it curves northward from the plaza, a serene casita, Quinta Maria, rests in a clearing, its raised porch facing south, its main entrance facing north. The riverbanks cross under Hildebrand Avenue and leave Miraflores for the Blue Hole, seeking the San Antonio River's originating spring. Thousands of years ago, tribes visiting the ancient river source designated it a sacred space. The Sisters of the Incarnate Word and their students sometimes navigate canoes along this calm stretch.

In the distance, off to the east, the Calzada del 2 de Abril drive curves from the main gate at Hildebrand toward the small house, hinting of a journey complete. Trees press in on every side of grassy expanses of poppies, roses, wild herbs, and flowering fruit trees, quiet and secluded.

54 North bridge
55 Quinta Maria
56 Mother and Child
57 Palapa bench
58 Pedestrian gate and staircase
59 Saguaro cactus
60 Winged Victory of Samothrace, with lions
61 Banca de Hernán Cortés
62 Quatrefoil pool

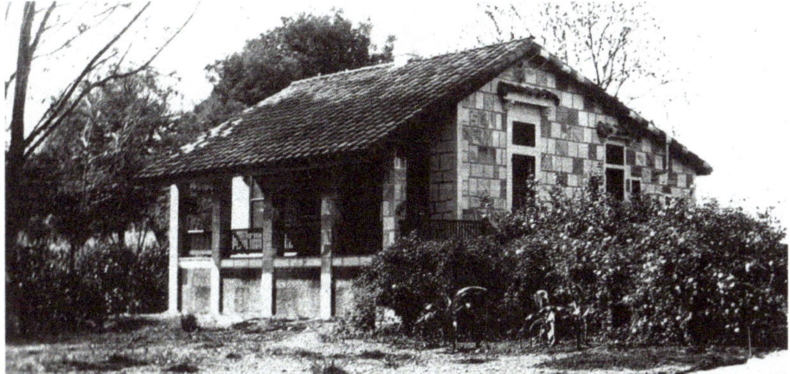

South and east facades of Quinta Maria, 36 x 28 feet, now in ruins.

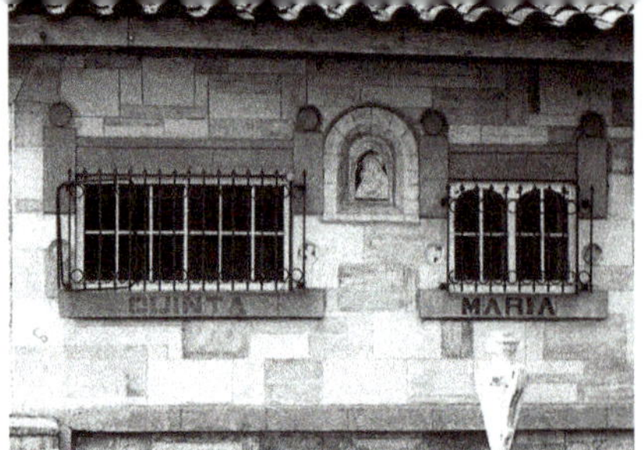

Quinta Maria, north facade.

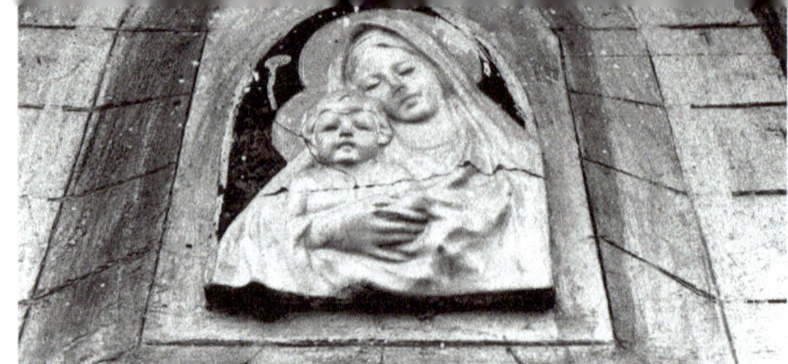

Mother and Child, 16 x 12 inches, in disrepair.

Quinta Maria: A Modest Casita

Quinta Maria, modest with a veneer of warm reddish brown or grayish stone, contrasts starkly with the doctor's grand home on Broadway. »Each facade of the casita has its own personality, somewhat eclectic though quaint in style. Too many steps, doorways, and windows abound for such a small cottage, making it accessible from almost every side. An iron-railed balcony looks west, a small north bridge crossing the river in view. A few modest sleeping spaces, a kitchen, and a bathroom cluster together in the rectangular building.«

The versatile house serves as a cooling retreat from San Antonio's summer heat, as a busy preparation area to stage large gatherings, and as a convenient respite for family members transitioning to a new or remodeled home.

As usual, Urrutia names the edifice. The words "Quinta Maria" run in blue tile inlays across the northern sills of two windows. Between them a white porcelain "Mother and Child" relief peers out from its own arched window toward the passing stream.

The Mother and Child relief connects Miraflores with its neighbor just upstream, the Sisters of Charity of the Incarnate Word, which shares with Urrutia a mission of healing and teaching. The sisters serve as nurses at Santa Rosa Hospital, where he often performs surgery, and he serves as a teacher for their nursing students and young doctors. Urrutia often attends a pre-sunrise morning mass across the street at the Incarnate Word Chapel before heading downtown to the hospital. The sisters wait for him at the chapel door and open it as he approaches.

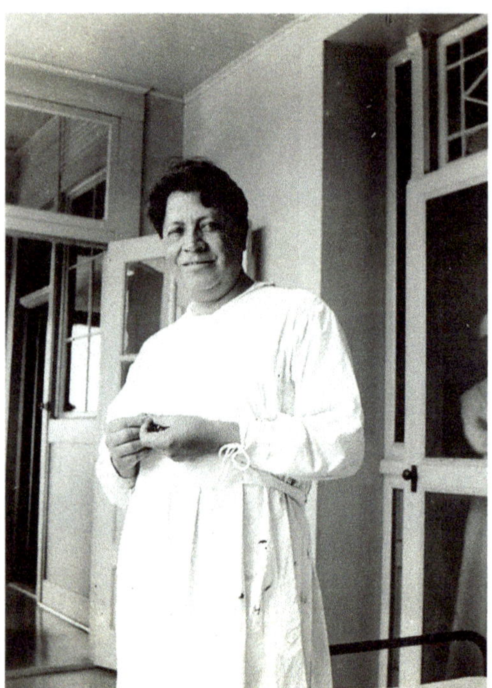

Urrutia at the Santa Rosa Infirmary, ca. 1920.

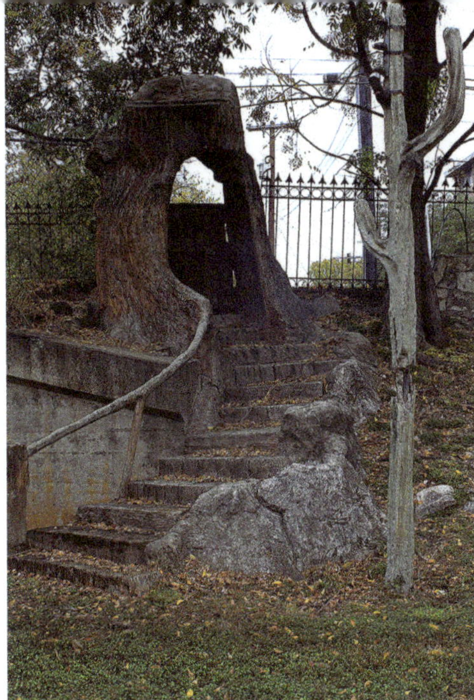

Hollow-tree pedestrian gate and staircase with saguaro cactus, *trabajo rústico*, 20 x 9 x 20 feet and 12 x 2 x 0.5 feet, respectively, since restored.

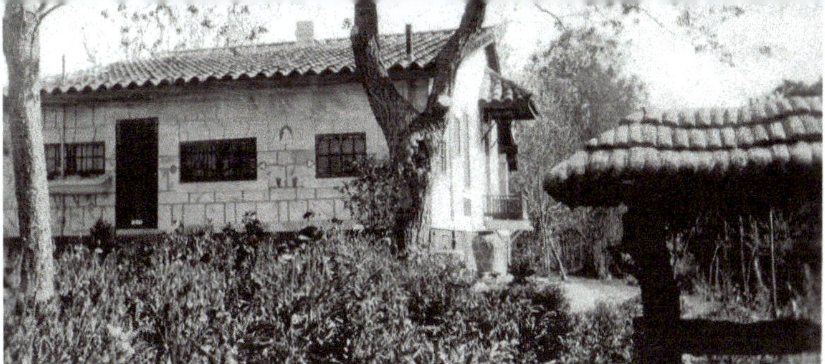

Palapa bench (partial view), *trabajo rústico*, 9 x 10 x 5 feet; with Quinta Maria.

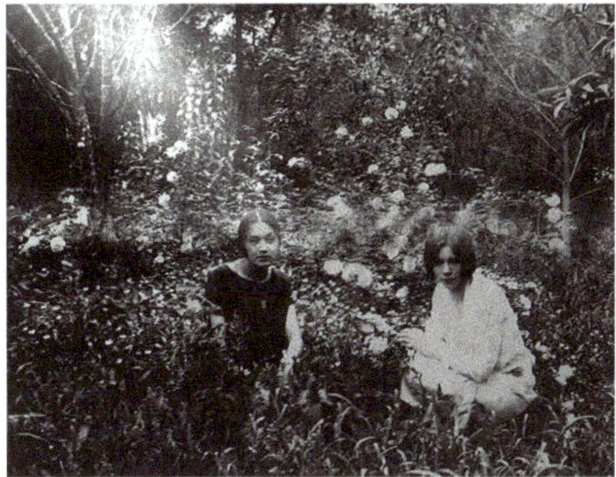

Urrutia daughters, probably Luz and Alicia, ca. 1921.

A Fairy Tale Wilderness: An Enchanted Moment

»The air around Quinta Maria smells of cottage garden herbs, peach trees, and shrub roses. A romantic palapa bench of faux bentwood logs and planks with a thatched roof rests nearby.[137] It sits along the bank at the edge of a swaying poppy patch where the gurgling river enters the property. Heavily wooded all around, the meadowy clearing north and east of the casita reveals an enchanting panorama that is quintessentially Urrutia.« An elaborate *trabajo rústico* pedestrian gate, as if fashioned from a centuries-old hollow tree, crowns the high wall above, rolling a dozen plank-like steps downward into the garden from Hildebrand. In some years, engulfed by overgrowth, it waits to be discovered; in other years, it beckons visitors clear as day. At the foot, a lanky faux saguaro cactus reaches for the sky; its original use was to cleverly disguise the delivery of electrical wires to the property.

The Winged Victory, Nike of Samothrace

Nearby, at the center of the clearing, the goddess Nike stands, wings outstretched, robes flowing in an eternal headwind. Two crouching lions, mouths agape, guard her side. She seems poised for flight on her stepped platform—or perhaps she has just arrived.

In his final touch to the garden, Urrutia transports his replica of the Hellenistic force of Victory, the Winged Victory of Samothrace, from his Broadway house, Quinta Urrutia, to the garden a few years before the house is sold and demolished in 1962. Atop Quinta Urrutia, the statue faced slightly south toward Urrutia's homeland. There, Mexico's own golden Winged Victory has perched high above the capital's Avenida de la Reforma since 1910, the sculpture placed there by President Díaz to celebrate Mexico's centennial.[138] At Miraflores, in contrast, Nike faces solidly into Urrutia's adopted country, toward his neighbors at Incarnate Word.

The statue testifies to Urrutia's worldliness and his serious appreciation of art. Theories evolved about the impetus for the original rare masterpiece of the second century BCE, uncovered in Greece a few years before the doctor's birth. But Urrutia knew her as a tribute by the Macedonian city of Samothrace to a splendid naval victory. In 1884, the Louvre acquired her, placing her at the top of its Daru staircase.[139]

Urrutia undoubtedly saw the Winged Victory in Paris, but a replica also existed at the Academia de San Carlos, established in the mid-1700s and known as the first art school of the Americas.[140] The replica arrived in Mexico to serve as a model for art students.[141]

In 1920, on the occasion of Urrutia's forty-eighth birthday, his friend Nemesio García Naranjo spoke of the statue as a symbol for Urrutia and said it reminded him of the art of a ship's prow:

> On the front of the house, Doctor Urrutia put the Victory of Samothrace, that wonderful Victoria as d'Annunzio said, is dressed in air, and makes the ship to which it is attached move, just by moving its luminous wings. A beautiful symbol chosen by Urrutia to protect his home which presides over his entire existence; his soul, always tied to the mast of duty and work, carries forward his ship with its invincible beating of wings. . . . When a life has been fruitful like that of Urrutia it deserves to be celebrated for inspiring character and virtue.[142]

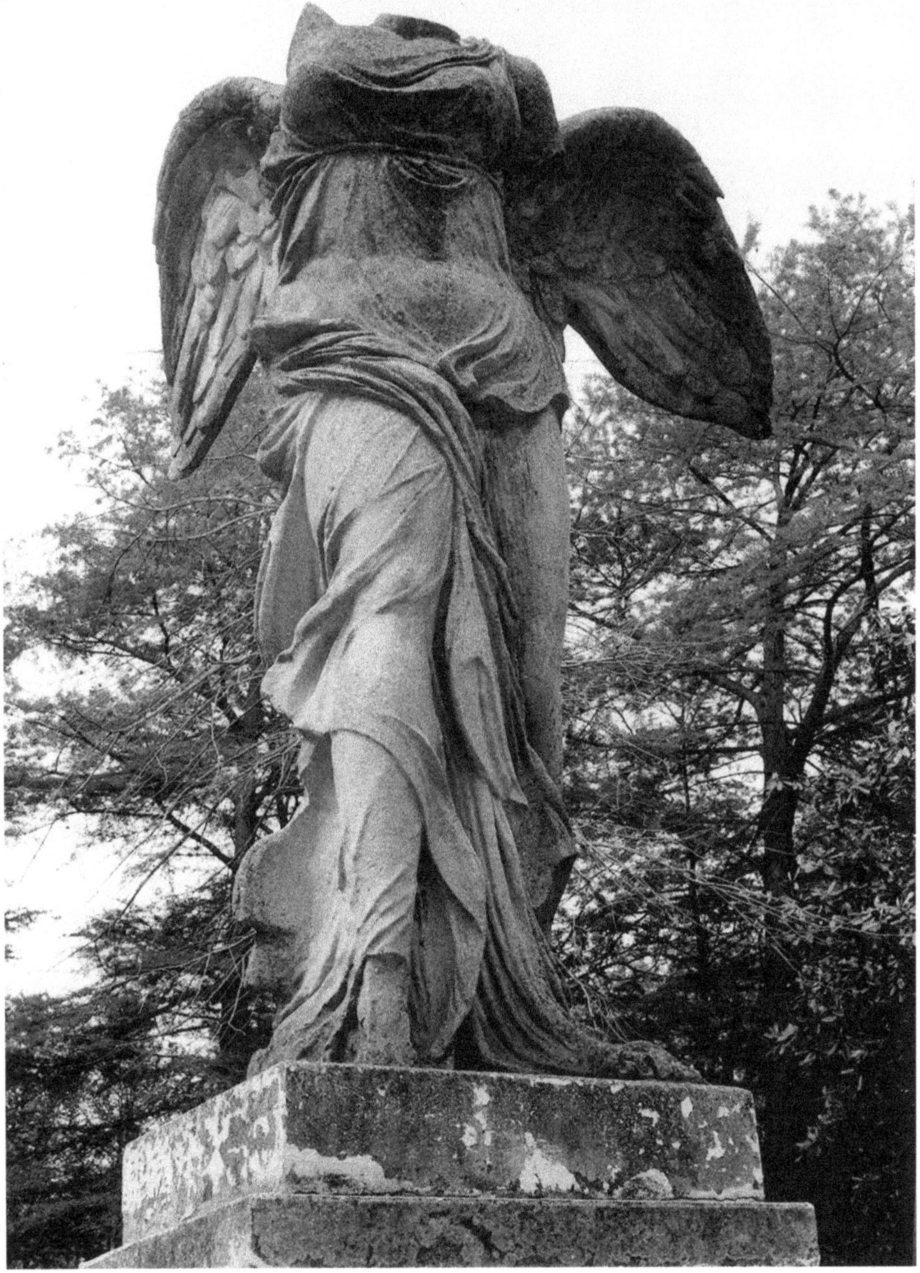

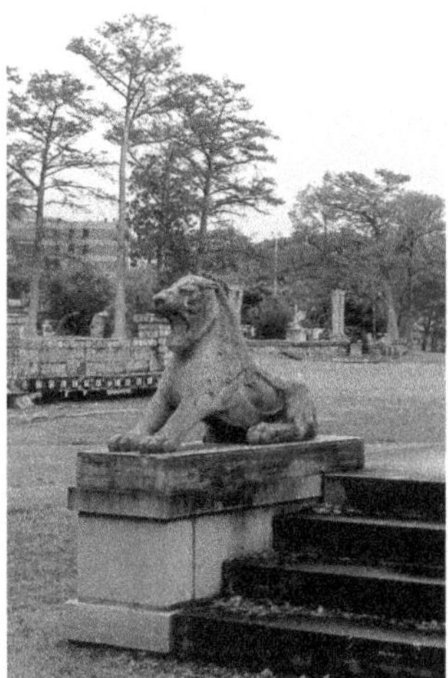

Lions, concrete, 3 x 2 x 5 feet.

Winged Victory of Samothrace, replica, concrete, 8 x 6 x 6 feet.

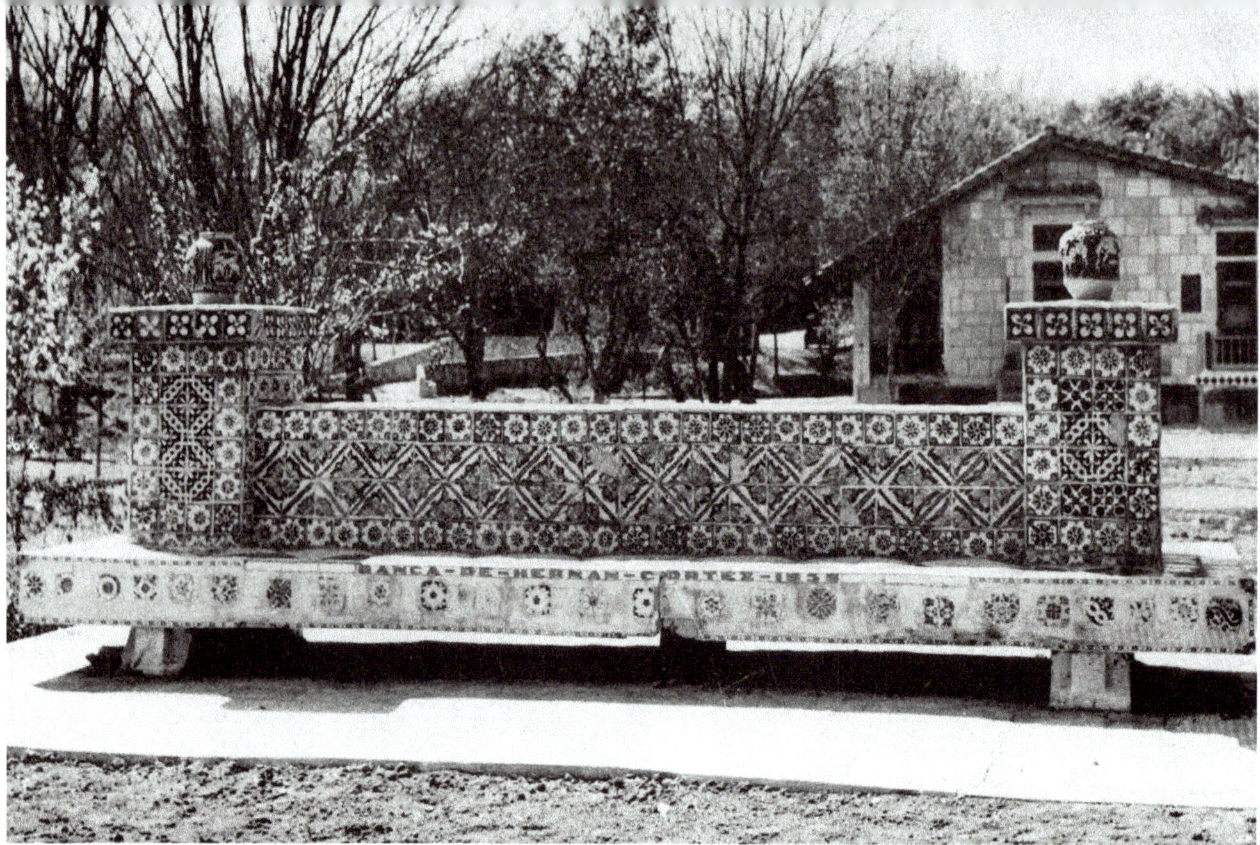

Banca de Hernán Cortés, 6 x 16 x 4 feet, now in ruins.

Banca de Hernán Cortés: An Ornate Return

Past the winged goddess, across the meadow from Quinta Maria, sits a magnificent Talavera bench created from more than seven hundred tiles. »Next to it lays a small quatrefoil pool near the Calzada drive.« Urrutia imagines that such a bench resided at the Cuernavacan palace of Hernán Cortés. The structure is inscribed with "Hernan Cortez [*sic*] 1535" on one side and "Doctor Urrutia" on the other. The bench links Urrutia with the conquistador four hundred years earlier. This garden of exile seems particularly significant here in San Antonio, once a territory of conquest, now a place of independence, freedom, safety, and reflection. The ornate bench offers an anchored respite, as if to say, "Sit here for a while, look deeply into the water, and think about where you've come from and where you will go."

Coda: "There and Back Again"

If it weren't for the cactus, the earthy Mexican style, and the warm subtropical breeze, the Quinta Maria area of the garden might evoke the storied traditions of C. S. Lewis, J. R. R. Tolkien, or the Brothers Grimm. But Mexico has its own fairy-tale tradition. Mexican children in the late nineteenth century, like Urrutia, knew these stories and relished the small illustrated newsprint booklets of them sold for pennies on the streets.[143] A stroll through Miraflores was not unlike a visit to the realms presented in those tales.

In the end, Miraflores travels back and forth in time and place, recalling—and in the process, reviving—stories and histories of specifically Mexican significance. Exile motivated its brilliant creator to use his avocation and resources to create in San Antonio a visual memory of a place to which he could never allow himself to return. In his garden, Mexico existed for him, keeping his memory of it alive and providing an expansive view of his complex and diverse homeland that he could share with others.

Women of the Garden

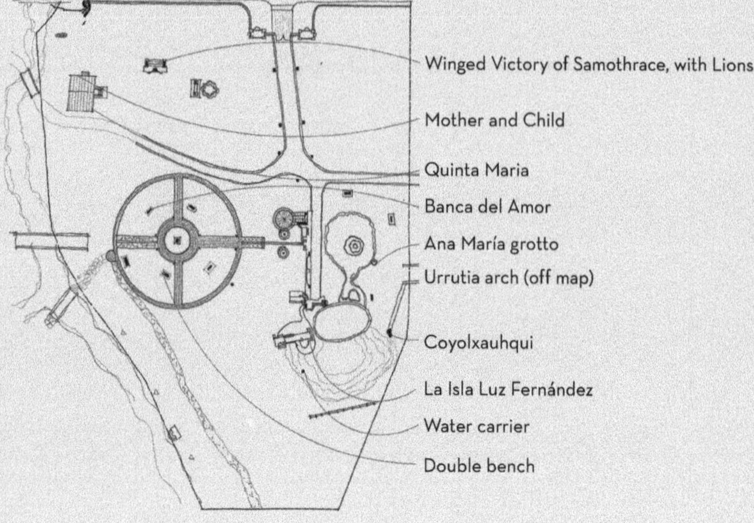

Urrutia placed many objects representing women throughout Miraflores, including the following:

- **Urrutia arch.** Displaying the Virgin of Guadalupe and the Mother and Child.
- **Ana María grotto.** Commemorating the birth of the eldest daughter of Urrutia's second wife.
- **Coyolxauhqui and Coatlicue.** Referencing feminine figures in the origin stories of the Aztec people.
- »**La Isla Luz Fernández.**« In memory of Urrutia's first wife.
- »**Water carrier.**« By Pantaleón Panduro.
- »**The rose courtyard.**« Again referencing Our Lady of Guadalupe.
- **Banca del Amor.** The Talavera bench with its tale of the Chinese princess.
- »**Double bench recalling the calif's wife and the Convent of Merced.**«
- **Mother and Child.** Relief on Quinta Maria.
- **Quinta Maria casita.**
- **Winged Victory of Samothrace.**

Urrutia's theme of women in the garden begins with Mexico's classical connection to the story of the Virgin of Guadalupe, a powerful influence in Mexican culture. But the importance of women in his own personal life cannot be overlooked. Urrutia's mother died when he was one, but he remembers her through a legend told by his family. At his first birthday celebration the pregnant mother single-handedly strangled a threatening lizard as she calmly nursed him under a shady tree. The next month, he says, "she pressed me to her heart and exhaled her last breath," passing away from complications of childbirth.

Of his four marriages, two lasted decades and were the bookends to his adult life. His first marriage, to Luz, lasted twenty-four years, until her death in 1921. His final marriage, to Carlota Gonzalez Miller, lasted twenty-five years, from 1950 until his death in 1975. Uncovering the lives of women from the nineteenth and early twentieth centuries presents challenges, but Luz was certainly an irreplaceable part of Urrutia's life. She was highly regarded in the Mexican and San Antonio communities, known for her grace and stalwart support of Urrutia during both their most successful and their most trying days. She bore eleven beautiful children who survived infancy, and she ran the household, dressed impeccably, and played the piano.

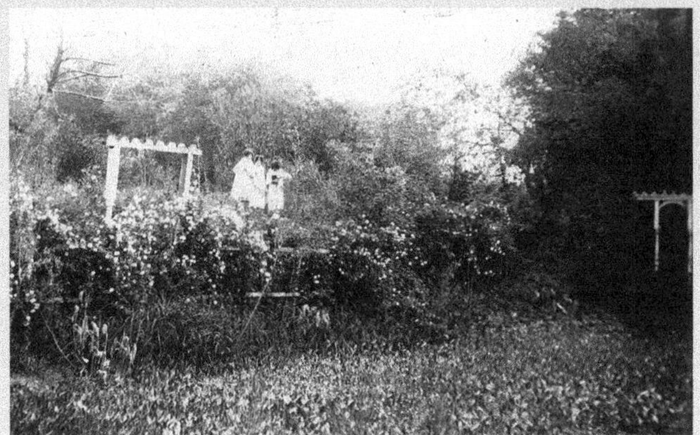

Urrutia daughters at Miraflores early in its development.

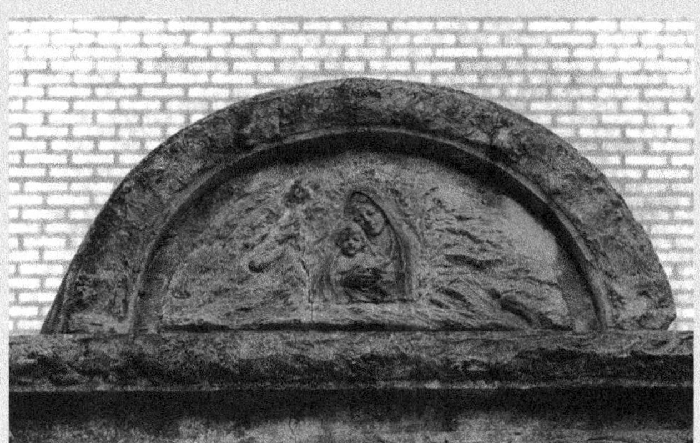

Urrutia arch, detail.

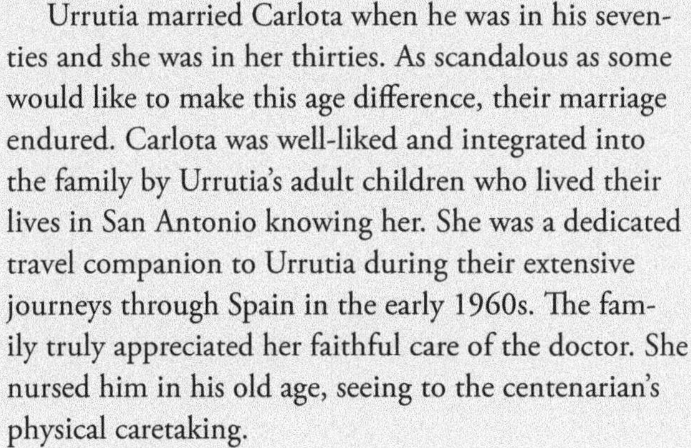

Urrutia married Carlota when he was in his seventies and she was in her thirties. As scandalous as some would like to make this age difference, their marriage endured. Carlota was well-liked and integrated into the family by Urrutia's adult children who lived their lives in San Antonio knowing her. She was a dedicated travel companion to Urrutia during their extensive journeys through Spain in the early 1960s. The family truly appreciated her faithful care of the doctor. She nursed him in his old age, seeing to the centenarian's physical caretaking.

The Urrutia household had a strong feminine presence in its early years in San Antonio. Ten of Urrutia's seventeen children were daughters. The two sons and two daughters of his second marriage were more distant, due to the marriage's failure and the children's relocating to Mexico at young ages. But Luz's seven daughters (one died a child in Mexico) grew up in the active household. Of particular note were the two eldest daughters, Refugio and Luz, ages eleven and twelve at the time of the escape from Mexico. Refugio became a pharmacist and worked alongside Urrutia. Luz, at age eighteen, upon her mother's death, assumed the duties of the busy household.[144] She never married. She handled the business end of the Farmacia Urrutia and kept the family archive, which Refugio assembled into a family volume after her sister's death in her early forties.

Three more daughters, Alicia, Dolores, and Maria Luisa, also worked in the pharmacy during their young adult years before marrying and moving away from San Antonio. Another daughter, Emma, married into the Paparelli family of San Antonio and helped run that family's pharmacy at historic Market Square nearby. The youngest daughter, Gloria, married a Mexican banker and repatriated to Mexico.

Another female presence, the Sisters of Charity of the Incarnate Word, occupied a special place in Urrutia's life. The Sisters served both as spiritual neighbors (they lived across Hildebrand from the garden) and as nurses, his assistants in his medical mission at Santa Rosa Hospital.

Urrutia also spoke on a number of occasions to women's groups about religion, gardens, philosophy, and other topics. It comes as no surprise that a value of women is visible at Miraflores, reflecting the strength of the feminine presence in Urrutia's life.

Epilogue

Aureliano Urrutia passed away in 1975 at the age of 103. Drawing on his indigenous roots, his knowledge of Mexican history, and his experiences of Mexico on the cusp of the twentieth century, he created a landscaped garden that captured his philosophy and reflected his Mexican heritage. Mexican artists such as L. L. Sanchez, Dionicio Rodríguez, and Marcelo Izaguirre helped Urrutia achieve his expression and contributed their commissioned work to the garden. Mexican writers such as José Juan Tablada, Nemesio García Naranjo, and Federico Allen Hinojosa also appreciated and supported Urrutia's efforts. And many other San Antonians, including Enrique Santibañez, Ignacio Lozano, Atlee B. Ayres, and Ray Lambert, as well as travelers from afar, visited Miraflores, thirsting to remember, reconnect, or learn something new of the Mexico that Urrutia knew and loved.

This magical place of winding paths and gravel roads, a wooded landscape punctuated by sculpture, ornate benches, and urns, and fantastical fountains and pools, fashioned into a metaphor for Mexico, provided a unique window into a different time and place. Unfortunately, the garden today is well on its way to disappearing.

People often ask, "Why is Miraflores in such a sorry state?" The short answer is that time and neglect have taken their toll on the garden since the mid-twentieth century. The various corporate owners of the property since its first sale in 1962 failed to study or recognize its importance and the relevance of its message. Of course, the simple answer belies a complex history.

Urrutia's Ownership, 1921–1962

Urrutia purchased the land for Miraflores from Guillermo and Margarita Alonso of Mexico City on May 5, 1921. The original boundaries of the fifteen-acre tract were Broadway on the east, the San Antonio River on the west, Hildebrand Avenue on the north, and two blocks south of Hildebrand on the south (if one draws a line from Allensworth Street at Broadway straight west to the river).[145] In 1953, Urrutia sold the "front" ten acres, the undeveloped area at Broadway and Hildebrand, to the United Services Automobile Association (USAA), retaining the remaining land, which included the area along the river that he had crafted into his five-acre garden.[146]

In 1962, at the age of ninety, after living in Quinta Maria at Miraflores with his wife, Carlota, for a few years, the physician transferred the garden to USAA. The deed stipulated that the place would be preserved "as a spot of beauty."[147] The property then passed along to two additional corporate owners—Southwestern Bell Telephone Company in 1974, which bought the entire reunited fifteen acres, and the University of the Incarnate Word in 2000, which received only the smaller 4.5-acre garden—before it was transferred to the City of San Antonio in 2006.[148]

Although Urrutia sold the property "with the condition of keeping the place like a spot of beauty, without touching any tree, anything that corresponds to this beauty," it has since suffered from both passive neglect and active destruction.[149]

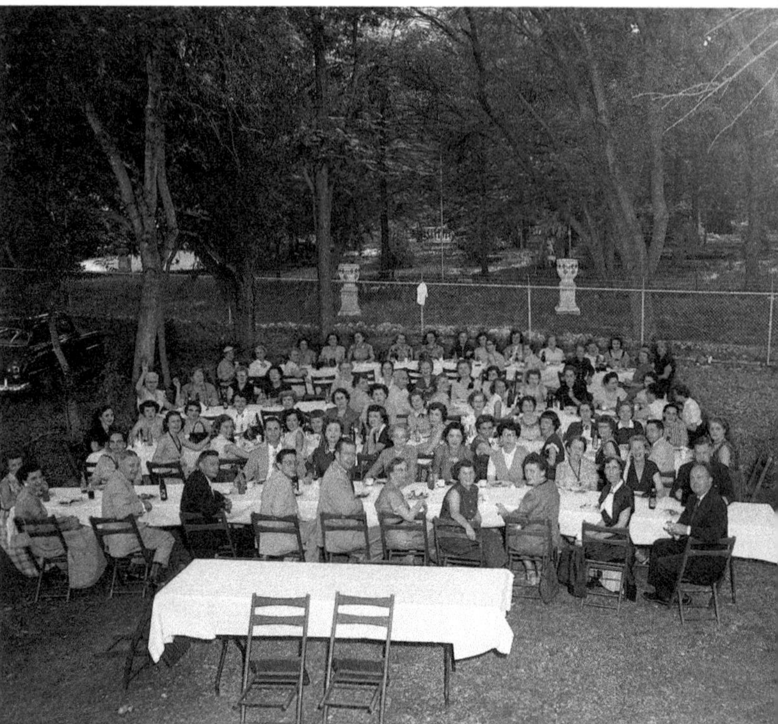

Looking west into Miraflores, at the western 1953 property line of USAA (the chain-link fence), showing the original level of forestation of the garden. USAA employees are seated in the foreground for an event.

USAA's Ownership, 1962–1974

During USAA's ownership of Miraflores, the company encroached on the 5-acre garden, converting it to expand their parking lot, and added a small building and a conventional rectangular swimming pool to the southwest quadrant as a recreational area for their employees' children.[150] The Urrutia arch was relocated from its original position along Broadway, between Hildebrand and Allensworth, to a location on Hildebrand near the northeast corner of the remaining 4.5-acre parcel.[151]

Southwestern Bell Telephone's Ownership, 1975–2000

In the mid-1970s, when Urrutia passed away, a number of journalists observed that the Miraflores property was in decline and disrepair. Although Southwestern Bell Telephone initially professed a desire to "continue the preservation" of the property, by 1977 the company had assigned responsibility for it to its volunteer organization, the Telephone Pioneers of America, which leased the property from the company.[152] Photographs from the late 1970s show an overgrown landscape in disarray, with no noticeable preservation effort, although a number of sculptures, benches, and other objects still existed.

In 1981, the telephone company applied for a demolition permit to raze Quinta Maria, sparking the first preservation effort in the twenty years since the garden passed into corporate hands. The city's Office of Historic Preservation asserted that because the building was the last remaining Urrutia structure, was part of a "unique sculpture memorial site," and was used as Urrutia's summer house, it should not be destroyed.[153] The San Antonio Conservation Society and city preservation staff met with company officials and effectively dissuaded the company from destroying the small house.[154]

The Quinta Maria exterior was minimally restored. Despite the fact that the house, in more recent times, had been reclad with a reddish-brown cut-stone veneer, the remodel reverted the appearance of the house to a look that appeared in a black and white photograph from the 1930s. The restored house was clad in gray concrete-board siding painted with a darker gray grid design to roughly suggest a cut stone

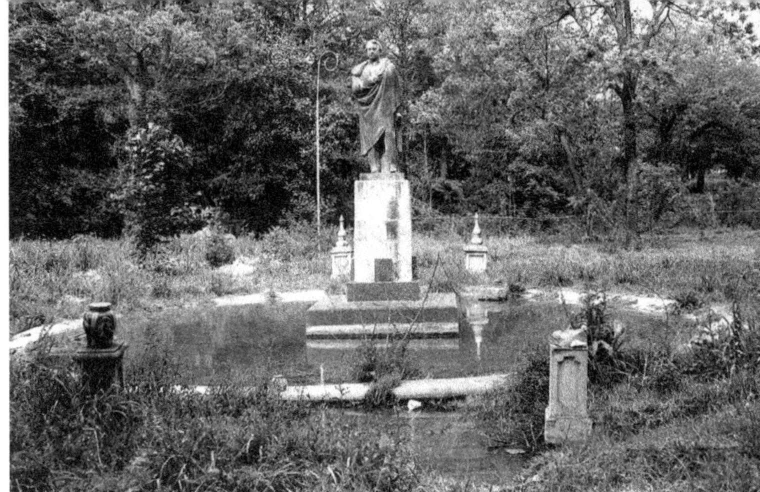

Reflecting pool during Southwestern Bell Telephone's ownership, 1978.

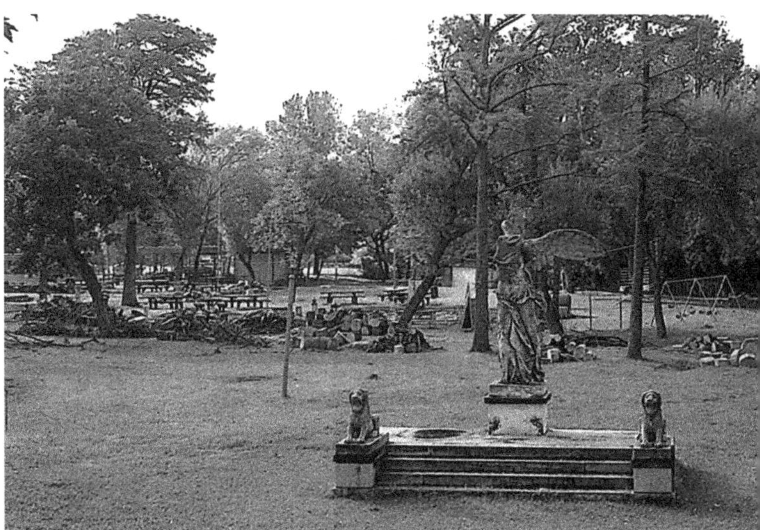

Garden during Southwestern Bell Telephone's ownership, ca. 1986.

surface. This siding remained on the casita at the time of writing. Although there was some discussion of using the cottage for a small telephone museum, the plan never developed, and for at least the next forty years, the structure remained in a state of decay.

During the remainder of the '80s, the telephone company Pioneers then cleared the land of trees without replacing or replanting the landscape. Tons of landfill were trucked in, underground pipes and drains installed, and original pathways destroyed or buried; the landscape's grade and appearance were significantly altered. Urrutia's aboveground swimming pool was cut in half, demolished, and hauled away or buried, and the large pond at the property's southern end was drained and filled in, along with other objects of the surrounding landscape. Winding pathways, native Mexican plantings, and the wooded landscape gave way to dozens of large concrete slabs bearing picnic tables and an approximately four-thousand-square-foot rustic pavilion with a tin roof, commercial barbecue pits, and bathrooms. The property was renamed Pioneer Park, and the Pioneers used it for company gatherings and activities.

Around this time, several important objects disappeared from the property. There were murmurs that Miraflores could be included in the Brackenridge Park master plan, but no action was taken.

In 1990, amid discussions about the importance of development along the river, the San Antonio River Authority investigated ownership and easements and concluded that a significant portion of Miraflores land along the river was city property. However, nothing further was pursued regarding these findings until more than a decade later.

In 1997, the telephone company gifted the Urrutia arch to the San Antonio Museum of Art. A group of citizens in favor of the move banded together to oversee the transfer. The museum stated an intention to restore the arch.[155] But more than twenty-three years later, at the time of writing this book, that work had not yet occurred. By the turn of the millennium, Miraflores was nearly forty years into a state of neglect, decay, and destruction.

Incarnate Word Ownership, 2000–2006

In 2000, the telephone company, renamed SBC, transferred ownership of the garden to the University of the Incarnate Word (UIW), and within a year the school had for unknown reasons destroyed the magnificent *trabajo rústico* Fuente Monumental, the garden's central fountain created by Dionicio Rodríguez.[156] Its destruction prompted a letter from the San Antonio Conservation Society requesting that the city take legal action and compel UIW to take responsibility for preserving Miraflores.

A few months later, in early 2002, the university planned to transfer fourteen of the garden's most precious objects, including the remaining works by Dionicio Rodríguez and L. L. Sanchez, to the San Antonio Botanical Garden, one mile away. This effort to divest Miraflores of significant artwork failed. Later that year, the university's public position became that it envisioned Miraflores as a recreation area and outdoor classroom and wanted to maintain some level of historic preservation. In the same breath, however, the university also planned to build thirty-eight parking spaces on the property. The city began to review the situation.[157]

In early 2003, the city filed suit against UIW, claiming title to a portion of the property, based on the 1990 San Antonio River Authority study.[158] In the resulting 2006 settlement, the city acquired all of Miraflores.[159] During the lawsuit, although Miraflores was nominated

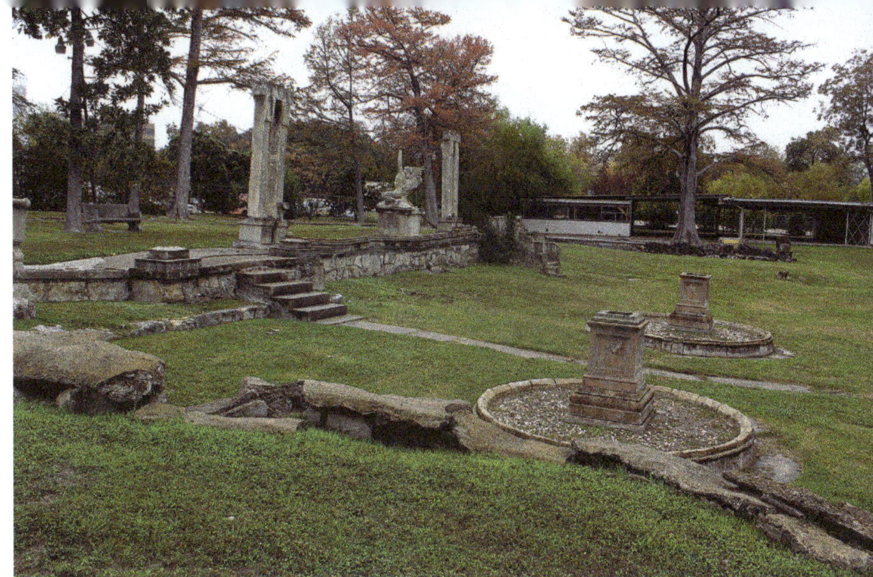

The current state of Miraflores, 2012. Standing where the Torre del Califa used to be, looking south. In the left half of the photo, you can see what is left of the esplanade—the two sets of doubled columns flanking the Cuauhtémoc statue. At the far southern end of the property is the pavilion built by Southwestern Bell. In the foreground at right are the two small pools that occupied the rose courtyard. The steps coming down from the esplanade at left have a sidewalk passing between the two small pools, which led to the Plaza del Centenario.

and approved as eligible for a National Register of Historic Places designation, the university as the owner opposed the nomination, blocking the designation.

City of San Antonio Ownership, 2006–Present

The 2006 lawsuit settlement provided funds for a master plan for Miraflores and development of a pedestrian bridge connecting Brackenridge Park to the property.[160] Under the new ownership, Miraflores was immediately added to the National Register of Historic Places. As part of the master plan, the University of Texas at San Antonio conducted an archaeological study of the area in 2008.[161] That same year, the city Parks Department stated that although it had no funds for future work, it would rely on a general fund and 2012 bond package for a phased funding approach.[162]

The pedestrian bridge was completed in 2009, the year the property received a State Archeological Landmark designation from the Texas Historical Commission, a further protection requiring review and permitting by the commission for future actions on the property.[163] Soon after, the city received an unsolicited proposal from UIW to manage and restore Miraflores, along with some of Brackenridge Park's northern historic areas. This raised concern from the San Antonio Conservation Society. The proposal was withdrawn.

In 2011, the conservation society raised $28,000 to restore the palapa bench and the Ana María grotto. It stated that "according to the 2008 Miraflores Master Plan, the projected cost for the redevelopment of the entire 4.5-acre park is estimated to cost over $8 million and will be funded by a municipal bond in 2012."[164] No money was specifically allocated to Miraflores in the city's 2012 bond package.

In 2012–13, a huge drainage project along Broadway and Hildebrand, the main corridors near the garden, diverted water flow underneath the northwestern corner of Miraflores to reach the San Antonio River. The project included digging up the corner of the property, dismantling the stone wall and wrought-iron fence built circa 1921, installing two six-foot-tall underground drainage culverts, and reshaping the way the land connects with the river in the area.[165] In exchange, with the support of the Texas Historical Commission and the San Antonio Conservation Society, the city agreed to stabilize the Hildebrand gate towers, replace and touch up the tilework on either side of the towers, restore the faux hollow tree pedestrian gate and staircase and saguaro cactus sculpture, and rebuild the landscape, the stone wall, and the wrought-iron fence affected by the intrusion.[166]

In 2016, the city approved funds of $229,000 to restore an original garden pathway, including landscaping and restoration of the statue of Urrutia. The pathway creates access from the pedestrian bridge into the plaza area, opening up the possibility of providing tours for educational and fundraising purposes.[167] Although the statue restoration was completed by the projected 2017 deadline, the walkway project proved challenging.

Just as the city was beginning to excavate the area, I was scrutinizing family photos and making multiple site visits to the garden in preparation for this book. My research revealed a different layout of the area being renovated than had been drawn in the city's plan, suggesting a grandness to the Plaza del Centenario that had not previously been understood. The city's plan, for example, featured a simple single brick walkway circling the pool. But my research showed a double ring of plaza brick walkways, with additional brick walkways bisecting the plaza from north to south and east to west, and the plantings and benches present in each quadrant that have been described in an earlier chapter.

I was able to collaborate with the city to help pinpoint the location of the original walkways, and the excavation uncovered the original brickwork and concrete, from under feet of infill in some places, almost completely in place. The excavation also revealed that the original Talavera benches located in the quadrants had been destroyed and buried. All this new information meant that new site plans and additional architectural drawings for the restoration had to be made. The garden as a whole, meanwhile, continues to decay as it awaits a long-term, cohesive solution.

What Does the Future Hold for Miraflores?

In 2019, the Brackenridge Park Conservancy, a nonprofit organization dedicated to protecting and preserving the 353-acre Brackenridge Park since 2009, signed a ten-year renewable agreement with the city to manage Miraflores and other park sites. The conservancy also commissioned a cultural landscape report that provides guidance on Brackenridge Park's significance and notes that Miraflores has great potential as a restoration candidate. It remains to be seen, however, what management of Miraflores by the conservancy will bring.

In celebration of the garden's centennial, Kathryn O'Rourke, a professor of art and art history at Trinity University, and I organized a half-day symposium in September 2021 to bring awareness to Miraflores's cultural significance. Along with landscape architect

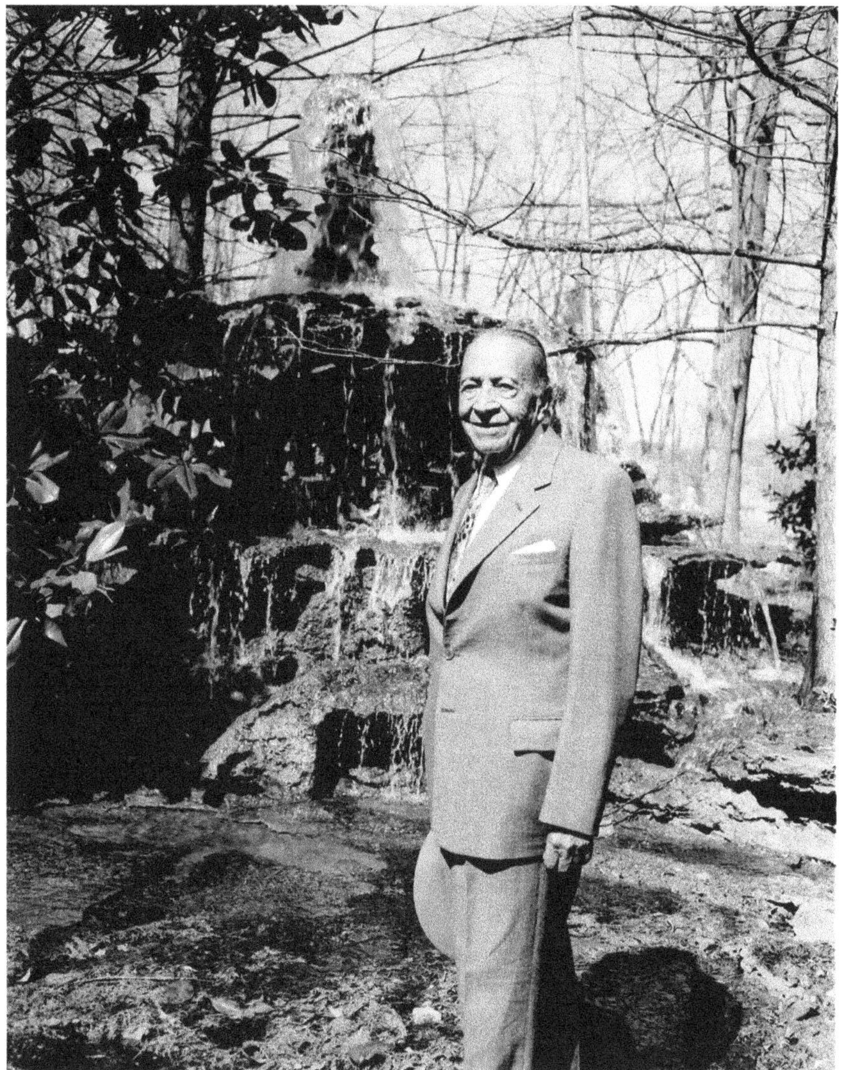

Urrutia at Miraflores, age eighty-seven, 1959.

John Troy; Jennifer Mathews, a professor of archaeology at Trinity University; and cultural scholar John Phillip Santos, we spoke to an audience of 150 about the garden's multilayered character. Speakers, property owners, neighbors, and others in the audience engaged in a discussion to consider the site's future. Scholar Tomás Ybarra-Frausto commented on the Mexican American community's relationship to the garden and expressed a hope for the garden's restoration and use as an educational resource. Yet even though the symposium was a broad-based show of support and concern for Miraflores, the garden's future remains tenuous and unresolved in 2022.

I hope this book will help Miraflores find new life. By exploring its messages, we can increase our understanding of Mexico and its place in our community—a wish my great-grandfather would have supported.

Miraflores was San Antonio's own Mexican garden, and even if it exists today only as a memory, there's no other place like it. The garden served as a special gathering place in San Antonio, and it provided solace, and at times shelter, for the Urrutia family after their exile from Mexico. With proper care and support, Miraflores could again offer the community a unique place where Mexican history and culture can be discovered, explored, understood, and celebrated.

Nos despedimos llevándonos la impresión de que el doctor Urrutia,
en su mexicanismo exaltado,
se envuelve inmaterialmente en una bandera tricolor.

We said goodbye with the impression that Dr. Urrutia,
in his exalted love for the culture and traditions of Mexico,
is ethereally wrapped in a tricolor flag.[168]

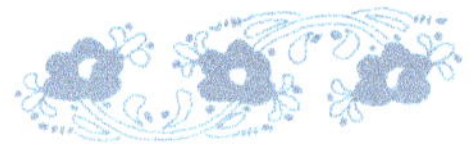

Appendix

The Plants of Miraflores

For a variety of reasons, but largely due to destructive environmental pressure or neglect, much of the planted landscape of Miraflores no longer exists. Photographs, writing, and memories, however, provide fairly detailed information about some of the plants Urrutia tended in the garden.

The northern sectors, especially the entrance sector through the Monumento gate at Hildebrand on either side of the Calzada del 2 de Abril drive, contained a forest of native pecan and elm trees, but the large Montezuma bald cypress, known in Mexico as the ahuehuete, dominated the garden—and for good reason. The tree holds a primary position in Aztec spiritual lore. The garden of Nezahualcoyotl at Texcoco contained a grove of a hundred of these trees, and Urrutia may have also planted a hundred of them. Chapultepec Park in Mexico City, also known as the Bosque de Chapultepec, originates from pre-Hispanic times as well, and contains a forest of ahuehuetes. Like Central Park in New York City, but two times larger, Chapultepec Park is often called the lungs of the city.

Urrutia wrote that he raised his ahuehuetes from seeds brought from Mexico. At the garden's southern end, a thick planting of ahuehuetes served as a backdrop to the view from the esplanade, across the pond. In front of them, a long white arbor traversed the view, punctuated with weeping willows and palms along the banks with the pond in the foreground. At the entrance to the Plaza del Centenario, fragrant magnolias introduced the garden room, and four ahuehuetes inside, one in each of the four beds, helped form the plaza space. Still more ahuehuetes punctuated the outer reaches of the plaza area, between the plaza and the esplanade. At the garden's western edge, along the riverbanks, Urrutia planted ahuehuetes, which survive today.

Other small fruit trees, including loquat and banana, presented themselves along walkways and paths, especially the gravel road between the plaza and the casita Quinta Maria. Bananas were located near the area of the pond and stage bridge, and a grove of peach trees stood near the southeast corner of Quinta Maria.

Medium-sized shrubs such as elephant ears and holly ferns arched along the pond banks and in proximity to other water features, and cacti clearly inhabited the rocky soil and the faux volcanic stones around the Fuente Monumental and rivulets that accompanied visitors along the esplanade.

Fragrant climbing roses clung to the esplanade's exterior raised walls, trailing from behind the Cuauhtémoc sculpture and culminating in an abundant display near El Baño de Nezahualcoyotl. Iris, water lily, and lotus populated the still pond waters below the esplanade and out into the pond's far edges. Queen Anne's lace grew in the wilder areas, with lily, iris, and canna occupying the garden's more cultivated areas. A field of poppies grew to the north and east of Quinta Maria.

Wild grasses provided transitions in some areas, while broad-leaved St. Augustine and Bermuda took hold in family gathering places in the sun or shade. Other areas included perennial herbs that Urrutia allowed to naturalize.

Urrutia claims that many of the plants had special origins, from varieties rescued from near extinction in Mexico to those with particular significance to him in his home environment. Gardening was clearly a serious avocation, born of his great love of the idea of the intersection of science and art.

TREES | Cottonwood (*Populus deltoides* var. *deltoides*), aka alamo; crepe myrtle; elm; live oak; magnolia; Montezuma bald cypress (*Taxodium mucronatum*), aka ahuehuete; palm; pecan, aka nueces, nogales; weeping willow

FRUIT TREES AND VINES | Banana, Carolina moonseed (*Cocculus carolinus*), loquat, mustang grape (*Vitis mustangensis*), peach

SHRUBS AND MEDIUM-SIZED PLANTS | Cacti, elephant ears, holly fern, nandina

FLOWERS | Canna, Castilian rose, iris, lotus, poppy, Queen Anne's lace, Turk's cap (*Malvaviscus arboreus* var. *drummondii*), water lily, other perennial flowers

HERBS | Perennial herbs for cooking and medicinal purposes

GRASSES | Bermuda, St. Augustine, wild grasses

Notes

1 Urrutia, *Pinacoteca.*
2 Urrutia Fernández and Urrutia Fernández, *Bodas de Oro*, 266.
3 Urrutia Martínez, Aureliano Urrutia, 37.
4 Urrutia, "Su Vida," 7–8.
5 Ibid., 11.
6 Ibid., 12–13.
7 The institution was cofounded by Luz Fernández and the archbishop of Mexico, José Mora y del Río.
8 Urrutia Fernández and Urrutia Fernández, *Bodas de Oro*, 11.
9 Urrutia, Sanatorio Urrutia.
10 Urrutia, "Su Vida," 13.
11 Ross, "Victoriano Huerta," 302.
12 Urrutia, "Su Vida," 13–14. Also, a detailed scholarly account of Urrutia's service in the Mexican government, and of his life in Mexico, can be found in Cristina Urrutia Martínez's biography of Urrutia.
13 Urrutia Fernández and Urrutia Fernández, *Bodas de Oro*, 39.
14 "Mob Besieges Hotel of Ex-Minister," *New York Times*, May 18, 1914.
15 One young daughter, Ernestina, died in Mexico in December 1913.
16 "Dr. Urrutia to Build," *San Antonio Express*, May 22, 1918.
17 "Artist Dreams of Four Centuries in Bit of Mexico Transplanted Here," *San Antonio Express*, December 20, 1918.
18 Bexar County Deed Records, book 638:202.
19 Aguirre and Urrutia, "Exiles of the Mexican Revolution," 51–56.
20 Tablada, *Las sombras largas*, 151.
21 Quoted in Urrutia Fernández and Urrutia Fernández, *Bodas de Oro*, 73, translated by the author. Original text: "*al florido sanatorio—alcázar de Coyoacán donde usted desarrolló paralela y armoniosamente su admirable obra de ciencia de belleza. Allí la sangre de las heridas que usted curó se transformaba en rosas en aquellas flores en que usted ponía sus más delicados y poéticos sentimientos de Xochimilco.*"
22 San Antonio map, 1908. City Engineers Office, San Antonio, Texas.
23 All three main elements of the arch have been displayed on the Uriarte workshop walls in Puebla, Mexico.
24 Hoar, "Miraflores: An Overview," 160. Hoar suggests the diverse influences on Urrutia's thinking at Miraflores.
25 Barber, *Maiolica of Mexico*, 96–97, image 56.
26 Urrutia practiced for nineteen years in Mexico and forty-six in San Antonio. At the time of writing, the City of San Antonio, the current owner of Miraflores, had removed these columns and some other relatively portable items and stored them.
27 Sierra, *Political Evolution*, 338–40.
28 Urrutia, "Su Vida," 6–7.
29 Ibid., 7–8.
30 The father of artist Frida Kahlo, Guillermo Kahlo was an accomplished photographer in his own right.
31 Urrutia Fernández and Urrutia Fernández, *Bodas de Oro*, 18; Urrutia Martínez, *Aureliano Urrutia*, 64–65.
32 Barber, *Maiolica of Mexico*, 3–10, 104–7. See also "Collecting the Arts of Mexico," Metropolitan Museum of Art.
33 Coote, *Eclectic Odyssey*, 26.
34 Ayres, *Mexican Architecture.*
35 Ayres & Ayres, Architects, Records, Alexander Architectural Archives. University of Texas Libraries, University of Texas at Austin.

36 Ibid.
37 Ibid.
38 Ibid.
39 Ibid. See also "The Use of Old Spanish and Mexican Tiles," *California Arts and Architecture*, 1932.
40 These large boulder-like formations are similar to those that artist Dionicio Rodríguez made for Chapultepec Park in Mexico City. See Light, *Capturing Nature*, 17–18.
41 The faux bois stump, along with a faux saguaro cactus, created by Dionicio Rodríguez, were two clever features designed to bring electricity into the garden unobtrusively.
42 Urrutia was well educated on the history of Spain and its influence on Mexico. He also became more aware of the Islamic influence on Spanish architecture through his work on Quinta Urrutia with Porfirio Treviño. He references the Mozarabic influence on architecture in his writing about Quinta Urrutia and his Talavera benches. Mozarabic architecture refers to the use of Islamic decorative motifs within Christian architecture after the Arab invasion into Spain. Urrutia traveled to Spain several times, where he saw Mozarabic architecture firsthand.
43 Urrutia Fernández and Urrutia Fernández, *Bodas de Oro*, 50; full text: "*Sabe Ud. lo que, al estarla contemplando, me dijo la 'Torre del Califa?' . . . Me confesó: Estoy aquí para dar fe de que mi dueño ha cumplido con magnificencia el precepto arábigo: 'Haz un hijo, planta un árbol, escribe un libro' . . . Como padre es un patriarca y les ha dado a sus numerosos vástagos. No sólo la vida, sino el ejemplo de su existencia admirable. Los árboles que plantó son selvas que aquí 'allá' don sombra y salud a los cuerpos y sereno gozo a los almas. En cuanto al libro escrito por su bisturí sin igual, tiene por páginas, no hojas de papel, sino vidas humanas por él arrancadas al dolor y a la muerte!*"
44 Urrutia, *Pinacoteca*, 101, translated by the author. Original text: "*Hecha con el muestrario de Talavera de la Reina, conservado como un tesoro del Siglo XVII. De la colección Olavarrieta, de Puebla. En este muestrario de platones y porcelanas se descubre la influencia mus-árabe en la industria Española y la influencia china en la industria Poblana.*"

Around the turn of the century, the well-known collection of Don Alejandro Ruiz Olavarrieta, from Puebla, was transferred to the Mexican government. Urrutia's friend, the artist Dr. Atl, culled and cataloged the collection. Urrutia probably viewed Olavarrieta's collection and may have collected some objects at that time. The government collection was initially housed by the Instituto Nacional de Bellas Artes; it is now housed by the Museo Nacional de San Carlos. Terry, *Terry's Mexico*, cxliii. Also "La Gran Colleción de Pinturas de Olavarrieta," *Arte y letras*, February 14, 1909, 12.
45 The large serving bowls are typical of Talavera polychrome pottery originating from the eighteenth century. The rooster panels, a known Uriarte design, were also used by Atlee B. Ayres in one of his Spanish Colonial Revival homes. The blue and white square tiles, bisected corner to corner, were often used as geometric borders or to cover large areas with a bold pattern.
46 Smithsonian Inventory, IAS TX000108.
47 Urrutia Martínez, *Aureliano Urrutia*, 24.
48 It was said that Catalina traveled to Mexico for the birth of each of her four children, but records show that the second child, Ana María, was born in San Antonio. This may explain the reason for this monument.
49 It is interesting to note that Catalina's move to Mexico essentially reinstated Urrutia's offspring in his homeland. Despite the fact that Catalina's children did not know their father well personally, they knew about him, and one spent some months visiting at Quinta Urrutia. Some of Urrutia's children, grandchildren, and great-grandchildren have made efforts to stay connected across the border. Both families became highly accomplished in a number of fields. One granddaughter, Cristina Urrutia Martínez, pursued thorough scholarly research of Urrutia, clarifying many of the details of his life in Mexico before 1914.
50 The name "Fuente Monumental" is taken from a reference by journalist Federico Allen Hinojosa.
51 The artesian well likely also fueled the garden's other water features.
52 Westkaemper, "Early Twentieth Century Influences," ch. 4, provides a landscape-oriented study of Miraflores, including the observation that "Dr. Urrutia included cascading water, spraying water, gently dripping water, water in shallow pools, deep pools, reflecting pools, and naturalistic stream like channels, all within view of the slow moving quiet water of the river." At the time of Westkaemper's master's thesis, the garden had long since ceased to function in this way.
53 Urrutia, "Su Vida," 5, translated by the author. Original text: "*Xochimilco . . . El lugar más hermoso del mundo; no había llegado la civilización que acabó con su belleza y los*

hombres no estaban en condiciones de estimarla. La laguna de Xochimilco, era recipiente de todas las aguas cristalinas y puras del deshielo de los majestuosos volcanes que coronan el Valle de México. El Ytazihuatl [sic]: mujer blanca; y el Popocatépetl: cerro que humea.

Las aguas de las nieves de éstos gigantescos volcanes, atraviesan enormes capas de arena, recorren subterráneamente toda la sierra y al llegar a la base de las montañas dan origen a veinte manantiales que con sus aguas cristalinas y puras forman el lago de Xochimilco. El más importante de éstos manantiales, tenía cién metros de diámetro, diez de profundidad y sus aguas al brotar eran de cristal de roca, que dibujaban todos los objetos con los colores y las reverberaciones del iris, espectáculo único en el mundo.

Las aguas de todos éstos manantiales, corrían y se perdían en miles y miles de canales, que rodeaban pequeñas islas maravillosamente cultivadas y que se llaman Chinampas."

Urrutia once referred to this fountain as La Fuente de Juventud, or the Fountain of Youth, perhaps because the legend references crystalline waters like those from the volcanoes. Urrutia, *Pinacoteca*, 95.

54 As transliterated from the Nahuatl language, there are a number of spellings for Cuauhtémoc, including the most conventional spelling, Cuauhtémoc. Urrutia consistently used an alternate spelling, Cuautémoc.

55 If one entered through the Urrutia arch at Broadway, the entry road would have also terminated in this area just across the rivulet stream, about seventy-five feet due east of Cuauhtémoc.

56 Nemesio García Naranjo, "La Protesta," *La Prensa*, December 4, 1921.

57 Smithsonian Inventory, IAS TX000124.

58 "Monumento a Cuauhtémoc," *La Jornada*, February 16, 2005. See also O'Rourke, "Artful Copies," 63–64.

59 García Naranjo, "La Protesta," translated by the author. Original text: "*la estatua entera, desde las plumas altivas del penacho, hasta los dedos encogidos de los pies, está llorando una injusticia, una de las injusticias mayores que registra la historia humana. Basta recordar las opulentas construcciones de Palenque, de Chichen Itzá, de Mitla y de Teotihuacán, para comprender que fue algo grandioso y divino lo que se destruyó. Y aunque los monumentos coloniales proclamen la gloria de una civilización superior, jamás constituirán una argumento convincente para la triste raza sacrificada. . . . Por eso el indio de mármol protesta: su puño crispado parece el de Cuauhtémoc increpando a los dioses y amenazando al cielo. . . .*

Por eso Urrutia ha hecho bien en colocar a la orilla de un lago el mármol conmovedor: las aguas puras recogen silenciosamente la protesta, para quizás mañana elevarse al cielo en forma de nube reivindicadora, que reviente en una tempestad de rayos justicieros."

García Naranjo, incidentally, was a writer for *La Prensa*, San Antonio's first Spanish-language daily newspaper, which was founded by Ignacio Lozano, who was also a Mexican exile.

Many artists have since depicted Cuauhtémoc, adding their own interpretations and developing his powerful symbolism, including José Clemente Orozco and David Alfaro Siqueiros. A 1946 study for the Siqueiros mural is held by the Blanton Museum of Art. In 1969, Texas artist Luis Jiménez created his own sculptural version of Cuauhtémoc, *Man on Fire*. The McNay Art Museum has a 1999 bronze casting.

60 Affron et al., *Paint the Revolution*, 335.

61 Dr. Atl used volcanoes in his early works. See, for example, his print *Symbol of Colima Volcano*, 1911–14. Affron et al., *Paint the Revolution*, 18. He was known for exploring the volcanoes near Xochimilco. The landscape painter José Maria Velasco also painted the volcanoes around the turn of the century and would have been familiar to Urrutia.

62 San Antonio directories from the 1920s list Sanchez, and a few doors down, a P. Treviño.

63 Gary Ware, "School of Snakes Blends Art of Ancient Indians," *Valley Morning Star*, November 24, 1991.

64 Burian, *Architecture and Cities*, 280.

65 For more about the Rodríguez family in San Antonio, see Joe Holley, "Family Carved Solid Lives in San Antonio," *Houston Chronicle*, August 25, 2013. The Rodríguez brothers, as far as the author knows, were not related to Dionicio Rodríguez.

66 Ware, "School of Snakes."

67 Mattei, "Bowie through the Years."

68 Pacho, "Baltazar Izaguirre Rojo."

69 Aureliano Urrutia, letter to Ray Lambert, June 13, 1924.

70 Patsy Light's definitive book on Rodríguez's career, *Capturing Nature: The Cement Sculpture of Dionicio Rodríguez*, documents the locations and time periods of Rodríguez's work. Her 2004–2006 application for the National Register

is probably the first documentation of his Miraflores artwork beyond Urrutia's own records.

71 Light, *Capturing Nature*, 23.

72 Ibid., 20–21, 87–88.

73 Barber, *Maiolica of Mexico*, 32–33.

74 Bárcena, "El Museo de Productor Industriales," 75, 94. See also Duclós Salinas, *Riches of Mexico*.

75 Hough, "Ancient Central and South American Pottery," 343–45.

76 Mason, "Guadalajara Pottery," 405–8.

77 Nelken, *Ignacio Asúnsolo*, 11–19.

78 Ibid., 65–68. See also Raquel Tibol, "Ignacio Asúnsolo, notable retratista," *Proceso*, October 27, 2012.

79 Hadman, "Aha Poet of the Month."

80 Mata, *José Juan Tablada*, 24–29.

81 Urrutia Fernández and Urrutia Fernández, *Bodas de Oro*, 144, translated by the author. Original text: "*Aureliano, estos versos los hice antes de conocerte, guarda el original porque tengo la seguridad de que tu espíritu desde entonces se empeñaba en dirigir mi vida.*"

82 Nahuatl for "place of coyotes," the headquarters of the Spanish conquest of the Aztec Empire.

83 An area near Mexico City; also, a lava field in general.

84 This neighborhood near Coyoacán was the original location of the Franciscan monastery and the site of an 1847 battle (Mexican-American War) where the US Army defeated Santa Anna only miles from Mexico City.

85 A volcano south of Mexico City.

86 Nahuatl for "close to water"; refers to the ancient core of Mexico.

87 Nahuatl for "emerald stone"; an area in the northwest of Zacatecas.

88 A nickname also referencing Anahuac.

89 Nahuatl for "flower field"; refers to an area for flowers and crops; birthplace of Urrutia.

90 Urrutia Fernández and Urrutia Fernández, *Bodas de Oro*, 143, translated by the author. (A different version of this poem can be found in *Obras I: Poesía* [Mexico: Universidad Nacional Autónoma de México, 1971].)

Nueces de Miraflores, sois semillas
Y así hacéis germinar en mi conciencia
Arboles rumorosos . . . los recuerdos
¡De la memoria en la profunda selva!
Nobles árboles son a cuya sombra
Llénanse de añoranzas las praderas
De la nostalgia como agrestes flores
Al Sol abriendo sus corolas frescas.
Bajo esas frondas y sobre la grama
Danzan en ronda angélica
Ninfas y Musas—¡Ay!—las ilusiones
De nuestra juventud en Primavera . . .
¿Es aurora o crepúsculo? . . . Dorada
Surge de Coyoacán la añosa iglesia;
Las volcanes encumbran la blancura
De sus nieves eternas;
Extiende el Pedregal la marejada
De sus olas inmóviles y pétreas;
Churubusco revive de otros días
La trágica epopeya;
Safiro y lapislázuli remeda
Y mas allá, tras de los arenales,
Del Ajusco la brava serranía

Oasis al pie de la escarpada sierra,
Por sus rústicas góndolas surcada
De Anáhuac la Venecia;
Edén para románticos idilios;
Del Imperio lacustre última perla;
Chalchihuite entre plumas de esmeralda
Y del Valle de México la gema
Todo el azul del cielo Xochimilco
¡Recoje en sus lagunas de turquesa!
***
Nueces de Miraflores, el destino
Brotar os hizo en San Antonio Béjar
Cuando en la mano que os sembró, la pura
Sangre del Indio corre por las venas,
Cuando la previsión del que os plantara
Es en su celo idéntica
A la que fué formando las chinampas
Con puñados de tierra
Como los nidos de las golondrinas
Industriosas, alegres y alfareras . . .

Pues a no ser el Destino adverso
La misma mano previsora hubiera
Allá en el corazón de Xochimilco

Hecho brotar mejor vuestra belleza
O dando sombra al tianguis bullicioso
O en la plaza natal frente a la iglesia!
***
¿Pero qué importa? . . . Cual vuestros follajes
Hacia el azul nuestro ideal se eleva,
Dios es uno, no hay patrias en el cielo
Es nuestro el Sol, la Luna y las Estrellas
Los mismos que iluminan de la Patria
Los volcanes, los bosques, las florestas,
La misma luz de Dios que fiel retrata
Xochimilco en sus lagos de turquesa . . . .
***

91 Urrutia, *Pinacoteca*, 89–92, translated by the author. Original text: "*La ciencia podrá ser seca y enjuta como una semilla; pero hay que dejarla caer sobre los surcos con mano amorosa. Ese es el secreto del sembrador.*"

92 Ibid., translated by the author. Original text: "*Tiene conocimientos profundos en ciencias naturales y es un apasionado de las Bellas Artes. En cuadro de vivos colores lo conmueve y se enternece con una delicada melodía. Por eso en México fraternizaba con artistas y en el destierro sigue cultivando el trato amable de los poetas.*"

93 Aguirre and Urrutia, "Exiles of the Mexican Revolution," 52.

94 Allen Hinojosa, translated by the author. Original text: "*Abierta de par en par la hermosa reja de hierro forjado que constituye la puerta principal de Miraflores, penetramos en el magnífico jardín que cultiva con tanto amor el doctor Urrutia, al final del parque Brackenridge, entre Broadway y Hildebrand. Nos esperaba. Un vigoroso mastín, echado a sus pies, se levanta y trata de venir hacia nosotros. Lo llama él y regresa adonde está su dueño, en tanto que nos aproximamos, seguros de que no tendremos nada qué temer. El doctor Urrutia está solo; sentado en una de las muchas bancas de azulejos que hay en Miraflores, dando frente a la fuente monumental y la espalda a la torre del Califa. El crepúsculo pone un tinte esplendoroso en el paisaje admirable de aquel sitio encantador. La víspera ha llovido torrencialmente y mientras en todas partes de la ciudad no ha quedado de ello la menor huella, debido a la tremenda sequía de varios meses, en Miraflores parece que la lluvia apenas si ha dejado de caer. Traspuesta la ancha puerta y penetrando bajo la sombra de los nogales, ahuehuetes y álamos, hemos dejado de sentir el calor sofocante que envuelve a la ciudad. Dijérase que, en un instante, hemos abandonado a San Antonio para entrar en un paraíso.*

Y mira fijamente los altos ahuehuetes que se mecen graciosamente agitados por la brisa del atardecer, y en cuyas copas brillan, con los últimos fulgores del Sol, las perlas que dejó en ellas la lluvia de la noche anterior."

95 Ibid., translated by the author. Original text: "*Nos sorprende el doctor Urrutia mirando fijamente hacia el fondo del estanque, donde hay unos lirios primorosos. Es un milagro de color que emerge del agua tranquila y que pone en la retina una maravilla. Parece molestarle esto; quizás ha creído que no hemos estado atentos a su discurso y que hemos perdido sus palabras, en las que puso energía y sinceridad. Pero con esa rapidez de pensamiento innata en él y que es una expresión de su sentido artístico y de su amor por las flores, me dice, poniéndome su mano sobre mi hombro: '—¡Qué hermosos!, ¿verdad? Pues más le impresionarán al saber que estos lirios son de los que trajo Maximiliano a México. Son del Nilo, pero Maximiliano, que amó tanto a México, quiso traer lo mejor para su nueva patria. Yo traje algunos, sacándolos del olvido y del abandono en que estaban en Cuernavaca, y aquí los tiene usted. ¡Acaso allá, donde él mandó plantarlos, ya no existan!'*

Se levanta el doctor y me invita a dar un paseo por entre las callecillas del jardín. Manda traer después un refrigerio, que saboreamos a medida que la tarde esplendorosa se esfuma suavemente."

96 After Urrutia sold the property, a corporate owner changed the terrain with tons of infill, and it no longer retains its original slope.

97 At least two of Urrutia's grandchildren have vivid memories of their grandfather teaching them to swim here. Urrutia was known to swim even one winter when the waters were covered with a thin sheet of ice. Bud Urrutia and Rosemary Leon, conversations with the author.

98 This bench is thought to be the first of many of this design by Rodríguez. Light, *Capturing Nature*, 11.

99 Patrizia Granziera's excellent essay is the source of many of the facts regarding the garden of Nezahualcoyotl cited in this paragraph. Granziera, "Concept of the Garden," 185–213.

100 Urrutia Fernández and Urrutia Fernández, *Bodas de Oro*, 209.

101 The cat's face references a vessel uncovered in 1901 and housed in the Museo Nacional in Mexico City. In the early

1900s, the museum was known as the Museo Nacional Mexicano. It was renamed Museo Nacional de Arqueología, Historia y Etnografía in 1910 and Museo Nacional de Antropología in 1940.

102 Aguilar-Moreno, *Handbook to Life*, 246.

103 Like many Aztec sites, Nezahualcoyotl's garden was not mapped by archeologists until the late 1970s, after Urrutia's death. See Aguilar-Moreno, *Handbook to Life*, 246.

104 Urrutia Fernández and Urrutia Fernández, *Bodas de Oro*, 50, translated by the author. Original text: "*Ud. es un hijo legítimo de Netzahualcóyotl, quien hizo de Texcoco el más admirable centro cívico indígena y del regio poeta heredó Ud. el don poético que lo hace crear la belleza en torno suyo y hacer poemas, himnos, no con elementos literarios, sino con la tierra, el agua, los árboles y las flores!*"

105 O'Rourke, "Artful Copies," 63.

106 Aka Luis L. Sanchez.

107 Archeologists were excited to uncover another version of Coyolxauhqui in 1978. The large, oval, disc-shaped, high-relief carving of the goddess is perhaps the more recognized version today. Nicholson, *New Tenochtitlan Templo Mayor*, 79–86; Pasztorsky, *Thinking with Things*, 219.

108 Nicholson, *New Tenochtitlan Templo Mayor*, 82–83.

109 Thanks to Jennifer Mathews, an archeology professor at Trinity University, for mentioning that a few elements at the base suggest the earth goddess Coatlicue.

110 Aguilar-Moreno, *Handbook to Life*, 190–91.

111 Nemesio García Naranjo, "Tres Coronas," *La Prensa*, April 7, 1921, translated by the author. Original text: "*Fué la compañera fiel, la colaboradora incansable del marido, el puntal amoroso que sostenía los muros cuando vacilaban . . . ¡Oración reconfortante en el dolor, nube de incienso en el altar, campana de gloria en el Te deum! . . . Y entretanto, ella, siempre a su lado, como el hilo de cristal que le da riego a la encina, como la arboleda piadosa que le brinda sombra al cafetal.*"

112 Urrutia Fernández and Urrutia Fernández, *Bodas de Oro*, 142.

113 Water flow in this area appears to correspond roughly with earlier maps of the Upper Madre ditch, a waterway in early San Antonio.

114 Ayres & Ayres, Architects, Records, Alexander Architectural Archives. University of Texas Libraries, University of Texas at Austin, translated by the author. Original text: "*Estanque lleno de Poesía con caídas de agua y vegetación traída especialmente de Texcoco y Xochimilco. El Puente y la Isla que dan acceso al Estanque tienen una hermosa plantación de Ahuehuetes.*"

115 Urrutia, "Su Vida," 3.

116 Ibid., 2–3.

117 Ibid., 2, translated by the author. Original text: "*Al hacer una excavación, para construir mi alberca; encontré una gran piedra que dice: Año 1716: Aquí se celebró la primera misa.*"

118 Urrutia Fernández and Urrutia Fernández, *Bodas de Oro*, 105.

119 The pools were added in ca. 1940 with the renovation of the plaza area. Earlier photos show the courtyard with no pools.

120 As described by Urrutia.

121 At the time of writing, the City of San Antonio had drafted a plan to restore the walkways and garden beds of this area in an interpretive style.

122 De Gante invented the first open-air sanctuary in Mexico City. Used for the purpose of conversion to Catholicism, the space accommodated thousands of congregants more accustomed to outdoor ritual. See De la Maza, "Friar Pedro de Gante," 104–7.

123 "The Use of Old Spanish and Mexican Tiles," *California Arts and Architecture*, 1932. Urrutia's version of this Mexican folktale attempts to explain the oriental influence on the pottery and fashion of Puebla.

124 Ibid. Consistent with the reference to the Moorish ruler, the Convent of Merced is of Spanish-Moorish design.

125 Smithsonian Inventory, IAS TX000105.

126 Urrutia Fernández and Urrutia Fernández, *Bodas de Oro*, 245.

127 In 2017, the statue, after being cleaned and restored, was rotated 180 degrees to face east toward a 2007 pedestrian bridge constructed by the City of San Antonio from the neighboring Brackenridge Park.

128 Translated by the author. Original text: "*Esta estatua representa al sabio Dr. y eminente cirujano Don Aureliano Urrutia y fue hecha en México en el año de MCMXL por el escultor Ygnacio Asúnsolo.*"

129 The eldest son, Aureliano Jr., attended school in Mexico City and devoted his career to working at the Robert B. Green Hospital in San Antonio.

130 "Development Travels West on Houston Street," *San Antonio Express*, June 13, 1926; "S.A. Landmark Is Erased," *San Antonio Evening News*, August 25, 1971.

131 Urrutia Martinez, 236, citing Hugo Aréchiga and Luis Benítez, *Un Siglo de Ciencias de las Salud en México* (Mexico: Fondo de Estudios e Investigaciones del Conaculta y FCE, 2000).

132 Urrutia Fernández and Urrutia Fernández, *Bodas de Oro*, 237–40.

133 Ibid., 254–55.

134 "La Peste Blanca," *El Heraldo de México*, July 10, 1908; "El Adelanto Quirurgico en México," *El Imparcial*, November 18, 1909.

135 The bridge, created by Dionicio Rodríguez, resides in Brackenridge Park, across the river from Miraflores.

136 A number of people have mentioned a memory that there was a bridge in the garden south to southwest of the Brackenridge Park bridge built in 2007, but research is inconclusive.

137 Smithsonian Inventory, IAS TX000107.

138 O'Rourke, "Artful Copies," 55.

139 "Winged Victory of Samothrace."

140 "Making San Antonio the Second Mexico City," *San Antonio Express*, November 19, 1916. Also "Cape-Wearing Surgeon to Stay Here," *San Antonio Light*, July 5, 1938.

141 "Antigua Academia de San Carlos."

142 Urrutia Fernández and Urrutia Fernández, *Bodas de Oro*, 153, translated by the author. Original text: "*En el frontis de la casa, puso el Doctor Urrutia la Victoria de Samotracia, esa estupenda Victoria que como dijera d'Annunzio, está vestida de aire, y que hace avanzar la nave a la cual va atada, con solo mover sus alas luminosas. Bello símbolo el escogido por Urrutia para proteger su hogar y que preside toda su existencia, su alma, amarrada siempre al mástil del deber y del trabajo, lleva adelante su nave con sus formidables aleteos. . . . Cuando una vida ha sido fecunda como la de Urrutia merece celebrarse como estímulo para el carácter y la virtud.*"

143 López Casillas, *Posada and Manilla.*

144 It was not unusual in Mexican culture for an older daughter to serve in the household.

145 Bexar County Deed Records, books 638:202 and 634:268.

146 Ibid., book 3359:162. USAA built their headquarters, completed in 1956, on the property.

147 Ibid., book 4781:67.

148 Ibid., books 7455:145 and 8690:1052.

149 "Dr. Urrutia Sells Property," *San Antonio Express*, May 29, 1962.

150 Dunn, *USAA*, 371.

151 This occurred sometime around 1953 when USAA purchased the front ten acres of the property.

152 Andrea Wright, "Dr. Urrutia—Living Legend," *San Antonio Express-News*, September 22, 1974; Kay Dyer, "Garden Is All That Remains of Miraflores, Urrutia Estate," *North San Antonio Times*, December 29, 1977.

153 Anna Marie Pena, "City Seeks Preservation of Urrutia Summer House," *San Antonio Light*, December 16, 1981. See also "Bell Officials Plan Urrutia Meeting," *San Antonio Express*, December 17, 1981.

154 Pena, "City Seeks Preservation."

155 David Uhler, "Museum of Art to Be Home to Urrutia Gate," *San Antonio Express-News*, September 14, 1997.

156 At some point, according to the city, the artesian well that fueled the fountain was also permanently sealed.

157 Amy Dorsett, "Legal Tug-of-War May Be Played Out on Tiny Park," *San Antonio Express-News*, August 3, 2002.

158 *City of San Antonio v. University of the Incarnate Word*, Cause no. 2003-CI-10943.

159 Ibid., Cause no. 2003-CI-10943, Settlement Agreement.

160 The City of San Antonio approved a budget of $455,045.25 for the project. Ordinance 2008-11-20-1032.

161 Ulrich, Archeological Services Associated with Improvements.

162 Mary Miranda, "Sculpture Garden Fences to Be Temporary," *San Antonio Express-News*, August 20, 2008.

163 Macdougal, "Miraflores Awaits Restoration"; "Mira Flores [sic] Designated as a State Archeological Landmark."

164 Hinkson, "Miraflores Park."

165 City of San Antonio Ordinance 2010-09-30-0836.

166 San Antonio City Council, Municipal Plaza, March 31, 2011; City of San Antonio Ordinance 2011-03-31-0218.

167 City of San Antonio Ordinance 2016-12-15-0985 approving Phase IV funds.

168 "A México Espero un Maravilloso Porvenir," translated by the author.

Works Cited

Affron, Matthew, et al., editors. *Paint the Revolution: Mexican Modernism, 1910 –1950*. New Haven, CT: Yale University Press, 2016.

Aguilar-Moreno, Manuel. *Handbook to Life in the Aztec World.* New York: Oxford University Press, 2006.

Aguirre, Nancy, and Elise Urrutia. "Exiles of the Mexican Revolution in San Antonio." In *Three Hundred Years of San Antonio and Bexar County*, edited by Claudia Guerra, 51–56. San Antonio: Trinity University Press, 2018.

Allen Hinojosa, Federico. "Una Entrevista con el Doctor Aureliano Urrutia." *Hoy*, 1944.

"Antigua Academia de San Carlos, Museo y Punto de Referencia en el Centro Histórico." *Boletín UNAM-DGCS-717.* Ciudad Universitaria, Dirección General de Comunicación Social. December 14, 2015.

Ayres, Atlee B. *Mexican Architecture: Domestic, Civil, and Ecclesiastical.* New York: William Helburn, 1926.

Ayres & Ayres, Architects. Records. Alexander Architectural Archives, University of Texas Libraries, University of Texas at Austin.

Barber, Edwin Atlee. *The Maiolica of Mexico.* Philadelphia: Pennsylvania Museum and School of Industrial Art, 1908.

Bárcena, Mariano. "El Museo de Productor Industriales, Materias Primas y Minería, del Estado de Jalisco." *Reports and Documents* 69–70, Mexico Ministry of Development, Colonization, and Industry, 1891.

Burian, Edward. *The Architecture and Cities of Northern Mexico from Independence to the Present.* Austin: University of Texas Press, 2015.

City of San Antonio vs. University of the Incarnate Word. Cause no. 2003-CI-10943, Bexar County, Texas. Settlement Agreement. February 6, 2006.

City of San Antonio v. University of the Incarnate Word. Cause no. 2003-CI-10943. Filed February 2, 2003.

"Collecting the Arts of Mexico." Exhibition overview, July 17, 2015–September 4, 2017. Metropolitan Museum of Art. Accessed December 3, 2017, at www.metmuseum.org/exhibitions/listings/2015/collecting-the-arts-of-Mexico.

Coote, Robert James. *The Eclectic Odyssey of Atlee B. Ayres, Architect.* College Station: Texas A&M University Press, 2001.

De la Maza, Francisco. "Friar Pedro de Gante and the Open Air Chapel of San José de los Naturales." *Artes de México* 150 (1972): 104–7.

Duclós Salinas, Adolfo. *The Riches of Mexico and Its Institutions.* St. Louis: Nixon-Jones Printing Co., 1893.

Dunn, Edward C. *USAA: Life Story of a Business Cooperative.* New York: McGraw Hill, 1970.

Granziera, Patrizia. "Concept of the Garden in Pre-Hispanic Mexico." *Garden History* 29, no. 2 (Winter 2001): 185–213.

Hadman, Ty. "Aha Poet of the Month, José Juan Tablada." AHA Books, 2001. Accessed March 12, 2017, at www.ahapoetry.com/PP0301.htm.

Hinkson, Roberto G. "Miraflores Park." *Preservation Advocate* 48, no. 1 (Fall 2011): 6.

"Historic Centre of Mexico City and Xochimilco." UNESCO, World Heritage Center. Accessed December 16, 2017, at whc.unesco.org/en/list/412.

Hoar, William. "Miraflores: An Overview of Time, an Insight into a Man." *Studies in the History of Gardens and Designed Landscapes* 30, no. 2 (2010): 152–89.

Hough, Walter. "Ancient Central and South American Pottery, in the Columbian Historical Exposition at Madrid, in 1892." *United States Congressional Serial Set*, Issue 3322, 1895, University of California.

"La Gran Colleción de Pinturas de Olavarrieta." *Arte y letras*, February 14, 1909.

Light, Patsy Pittman. *Capturing Nature: The Cement Sculpture of Dionicio Rodríguez.* College Station: Texas A&M University Press, 2008.

López Casillas, Mercurio. *Posada and Manilla: Artistas del Cuento Mexicano.* Mexico DF: Editorial RM, 2013.

Macdougal, Bruce. "Miraflores Awaits Restoration." *Preservation Advocate* 46, no. 4 (Summer 2010): 2.

Mason, O. T. "The Guadalajara Pottery." *Science* 8 (1886).

Mata, Rodolfo. *José Juan Tablada: De Coyoacán a la Quinta Avenida.* México DF: Fundación para las Letras Mexicanas, UNAM, 2007.

Mattei, Eileen. "Bowie through the Years." *Experience HCISD Magazine.* Spring 2013.

"A México Espero un Maravilloso Porvenir, Afirma el Dr. Urrutia." *Excélsior*, July 26, 1929.

"Mira Flores [*sic*] Designated as a State Archeological Landmark." *Preservation Advocate* 45, no. 2 (Winter 2010): 9.

Nelken, Margarita. *Ignacio Asúnsolo.* México DF: Universidad Nacional Autónoma de México, 1962.

Nicholson, H. B. *The New Tenochtitlan Templo Mayor Coyolxauhqui-Chantico Monument.* Berlin: Ibero-Amerikanisches Institut, 1985.

O'Rourke, Kathryn E. "Artful Copies: The Urrutia Collection and Porfirian Visual Culture." *Mexican Studies* 30, no. 1 (Winter 2014): 31–70.

Pacho, Carlos Rubio. "Baltazar Izaguirre Rojo." *Enciclopedia de la literatura en México.* Ciudad de México, Secretaría de Cultura, December 10, 1997.

Pasztorsky, Esther. *Thinking with Things: Toward a New Vision of Art.* Austin: University of Texas Press, 2005.

Pomade, Rita. "Talavera-Mexico's Earthly Legacy from the City of Angels." *Mexconnect.* Accessed June 17, 2017, at www.mexconnect.com/articles/1061-talavera-México-s-earthly-legacy-from-the-city-of-angels.

Ross, Stanley R. "Victoriano Huerta: visto por su compadre." *Historia Mexicana.* México DF, El Colegio de México, vol. XII, no. 2, 1962.

Sierra, Justo. *The Political Evolution of the Mexican People.* Translated by Charles Ramsdell. Austin: University of Texas Press, 1969.

"Smithsonian Inventories of American Painting and Sculpture." Smithsonian American Art Museum. Accessed February 12, 2107, at https://americanart.si.edu/research/inventories.

Tablada, José Juan. *Las sombras largas.* Mexico: Consejo Nacional para la Cultura y las Artes, 1993.

Terry, T. Phillip. *Terry's Mexico.* London: Gay and Hancock, 1911.

Ulrich, Kristi M. Archaeological Services Associated with Improvements to Miraflores at Brackenridge Park, San Antonio, Bexar County, Texas. Center for Archaeological Research, University of Texas at San Antonio, Archaeological Report No. 387, 2008.

Urrutia, Aureliano. *Pinacoteca.* San Antonio: Self-published, 1940.

———. *Sanatorio Urrutia.* Coyoacán, DF: Bouligny & Schmidt Sucs., 1911.

———. "Su Vida y Su Origen." San Antonio: Self-published, 1959.

Urrutia Fernández, Refugio, and Luz Urrutia Fernández, editors. *Bodas de Oro del Doctor Aureliano Urrutia, 1895–1945.* San Antonio: Self-published, 1945.

Urrutia Martínez, Cristina. *Aureliano Urrutia: Del Crimen Político al Exilio.* México DF: Tiempo de Memoria Tusquets, 2008.

"The Use of Old Spanish and Mexican Tiles on Dr. Urretia's [*sic*] Estate, San Antonio, Texas." *California Arts and Architecture.* October 1932.

Westkaemper, Sarah C. "Early Twentieth Century Influences: San Antonio Gardening Style." Master's thesis, Louisiana State University, 1985.

"Winged Victory of Samothrace: A Closer Look at the Victory of Samothrace." Musée du Louvre. Accessed December 24, 2017, at musee.louvre.fr/oal/victoiredesamothrace/victoiredesamothrace_acc_en.html.

Image Credits

All photographs and illustrations are from the author's collection unless otherwise indicated.

pages xiv, xvi, 10, 20, 21 right, 22 bottom left and right, 24, 26 bottom, 37 bottom, 38, 41 left, 51, 52, 61 bottom, 63, 64, 71 left, 75 bottom right, 80 top left, 80 bottom right, 82 bottom, 84 right, 88 top left, 88 top right, 89 top right, 91 left, 91 right, 99 top, 101, photographed by the author.

page 2 bottom center, photographed by Schlatman Hermanos.

page 4 bottom left, photographed by M. Ramos.

page 6 right, Series title: Passenger Lists of Vessels Arriving at Galveston, Texas, 18961951. Record Group Title: Records of the Immigration and Naturalization Service, 17872004. National Archives at Washington, DC.

pages 7, 35 top, 103, reprinted by permission. Copyright © UTSA Special Collections / *San Antonio Express-News* / Zuma Press.

page 8, adapted by the author from a 1908 map of San Antonio.

pages 13, 14, 25 top, 32, 48 left, 58, 70, 72, 85 top, 86, 94, drawn by Rebecca Schenker, in collaboration with the author.

page 15, drawn by Rebecca Schenker, technical assistance by Benjamin Barenblat, in collaboration with the author.

page 16 top, Historic American Buildings Survey, National Park Service, Survey #TX-3558, 2011.

pages 18, 19 top, 95 right, photographed by Ansen Seale. Reproduced by permission. Courtesy of the San Antonio Museum of Art. Gift of Southwestern Bell and moved to the museum through the generosity of the Steves Foundation, 1997.

pages 25 bottom, 28 bottom, 39, 68 all, 80 top right, 80 center right, photographed by Atlee B. Ayres. Courtesy General Photograph Collection, UTSA Special Collections.

pages 26 top left, 29 top, reproduced by permission from the Ayres and Ayres Collection, the Alexander Architectural Archives, the University of Texas Libraries, the University of Texas at Austin.

page 29 bottom, reproduced by permission. Courtesy Louis D. Hamilton Collection, McNay Art Museum Library & Archives, San Antonio, Texas.

pages 30 top, 48 right, 98, 99 bottom, courtesy General Photograph Collection, UTSA Special Collections.

page 31, reproduced by permission. Courtesy *Arts & Architecture*, June 1932. Copyright © Travers Family Trust.

page 33, reproduced by permission. Copyright © *San Antonio Express-News* / Zuma Press.

pages 34, 67, 76, 78–79 top, photographed by H. L. Summerville.

page 44, photographed by Immanuel Fiedlaender. Reproduced via Wikimedia Commons from ETH-Bibliothek. Public domain.

page 45, reproduced by permission from Sylvia Sánchez-Pogson.

page 46, photographed by Mike E. Perez. Reproduced by permission.

page 47, photographed by Mónica del Arenal. Reproduced by permission.

page 49, photographed by Hugo Brehme, licensed under Creative Commons. Share Alike 3.0 Unported via Wikimedia Commons from Uriarte Talavera.

page 50, reprinted from *Science*, an illustrated journal, vol. 8, no. 196, 1886. Science Company, New York.

pages 62 left, 81 top, photographed by Kathryn O'Rourke. Reproduced by permission.

page 62 right, reproduced via WikiArt Visual Art Encyclopedia. Public domain.

page 65, photographed by H. L. Summerville, courtesy General Photograph Collection, UTSA Special Collections.

page 75 top, used with permission from DRT Collection at Texas A&M University–San Antonio.

page 75 bottom left, reprinted from Edwin Atlee Barber, *The Maiolica of Mexico* (Philadelphia: Pennsylvania Museum and School of Industrial Art, 1908).

page 81 bottom, photographed by Dennis Baltuskonis. Reproduced by permission.

page 88 bottom, reproduced by permission of the Sisters of Charity of the Incarnate Word Archives, San Antonio, Texas.

Acknowledgments

Many thanks to the many people who have supported me through this process, but most of all, thank you to:

My father, Dr. Aureliano A. Urrutia, for the many hours of conversation about the Urrutia family, and for giving me this legacy to explore and elaborate.

My husband, Mark Schlesinger, my partner, dedicated conversationalist, first reader, and much more.

My daughter, Noa Luz Barenblat; my son, Benjamin Ezra Barenblat; and my stepsons, Will and Sam Schlesinger.

The wonderful and vibrant Gwen Rhea Cowden.

Rebecca Schenker, for beautiful illustrations.

Karen A. Monsen and the late Dawn Bruner Finlayson, for advice, flexibility, and enthusiasm.

Kathryn O'Rourke and Nancy Aguirre—your insights, understanding, and writing about related topics have inspired me to explore my heritage, and your unselfish support and dialogue have enabled me to reach new heights.

Tomás Ybarra-Frausto, for your insights on my efforts to explore and communicate my family heritage and its significance in San Antonio, and for your generous contribution of the foreword. I am honored.

Tom Payton, Steffanie Mortis Stevens, and Sarah Nawrocki of Trinity University Press, for working with me to get this book off my desk and onto the shelf, along with Emily Jerman Schuster for your expert editorial hand.

Teachers, whose voices still inspire me: the late Ernest Mae Seaholm, the late Mary Louise Suggs Norman, Paul A. Foerster, and George Butte.

Many additional people helped me accomplish my work. Thank you to:

Robert Rivard, for publishing an initial in-depth article about Miraflores, and to *San Antonio Report* (aka *Rivard Report*) staff members at that time who so enthusiastically worked to publish my early articles as I began this project.

Rosemary Urrutia Leon, the late Bob Bobbitt, and Sister Consuelo Urrutia for your stories.

Mis amigos en México: el difunto Don Galdino Díaz Flores, Maestra Araceli Peralta Flores, su hijo César; artista y ingeniero Rodrigo Ocaña Lopez. Por un día involvidable y emocionante. ¡Hasta que nos encontremos de nuevo!

Robert Atlee Ayres; Sylvia Sánchez-Pogson; Mike E. Perez; Susan Toomey Frost; Donna Guerra of the Sisters of Charity of the Incarnate Word Archives; Heather Ferguson and Leslie C. Straus of the McNay Art Museum; Suzanne Weaver, Lana Meador, and Katie Spencer of the San Antonio Museum of Art; Katie Pierce Meyer of the University of Texas at Austin; Patsy Light; Sarah Reveley; Richard Halter; Luisa Bruno-Lopez; Beverly Bohacek; Mary Haley Eckstine; Dennis Baltuskonis; Gregg Eckhardt; Josh Huskin; and Char Miller, for generously sharing time and resources.

All those who toured the garden with me and provided thoughts and insights, in addition to Dad, Mark, Noa, Becky, Donna, and Kathryn, including Mónica del Arenal, Jennifer Mathews, Bill Pennell, John Troy, and John Phillip Santos.

Pat Jimenez, Vicki Yuan, Claudia Guerra, and Lynn Bobbitt, whose work has helped increase awareness of Miraflores in the community.

Mis parientes en México: A mi primo Walther Boelsterly, con gracias por nuestra visita en la Ciudad de México, tu correspondencia, tu esfuerzo por comunicarte conmigo a larga distancia, y por tus historias y tu perspectiva; a mis primos Fernando Urrutia y Evelyn Coufal Urrutia, por el día hermosa juntos, tus hospitalidad, y por tus historias de familia; y a mi prima Ximena Urrutia, por tu correspondencia y con los mejores deseos en tu filmación. ¡Hasta que nos encontremos de nuevo!

A very special thank you to Tom Shelton and Carlos Cortez of UTSA Special Collections, for your helpfulness and dedication; along with Nancy Sparrow and Beth Dodd of the University of Texas at Austin, Leslie Stapleton of Texas A&M University–San Antonio, Leslie Ochoa of the Witte Museum, Beth Standifird of the San Antonio Conservation Society, Nancy Hadley of the American Institute of Architects, and Jennifer Navarre of the Williams Research Center, for research assistance.

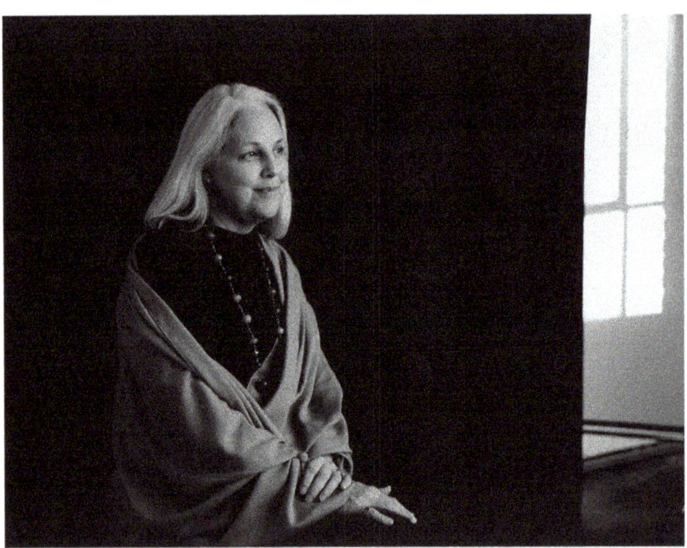

As a teenager, Anne Elise Urrutia first ventured into Miraflores to photograph the disappearing family garden of her great-grandfather, Aureliano Urrutia, a surgeon who immigrated to San Antonio from Mexico during the revolution. Over the years her research on Miraflores and her family history has allowed her to rebuild, through words and pictures, the doctor's lost landscape, uncovering his message of cultural heritage communicated through this once beautiful and expressive place. Urrutia received her English degree from Colorado College and blogs at quintaurrutia.com. She lives in San Antonio.

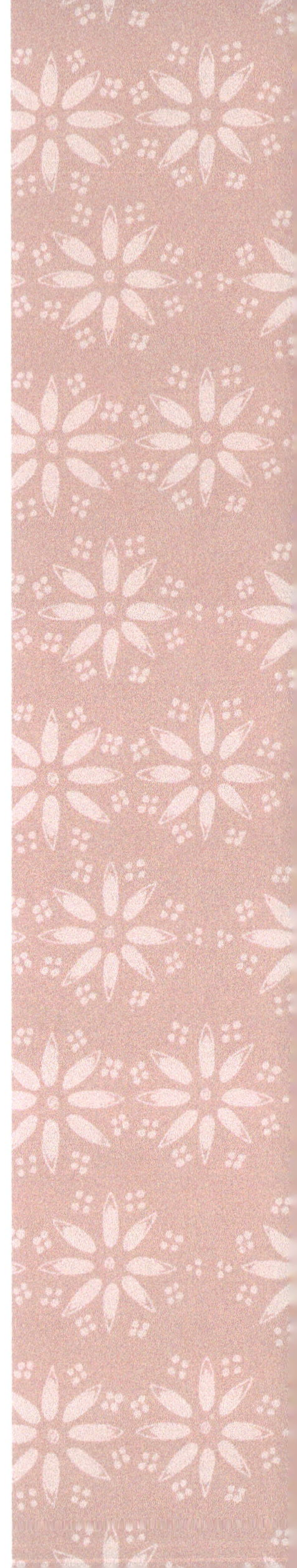

www.ingramcontent.com/pod-product-compliance
Lightning Source LLC
LaVergne TN
LVHW060618110826
845147LV00019B/1045

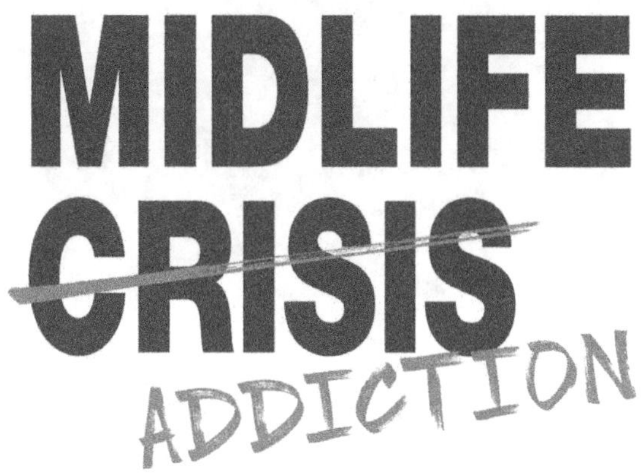

by
Carol Farley

Edited by Catherine Gigante-Brown
Cover and interior design by Vinnie Corbo

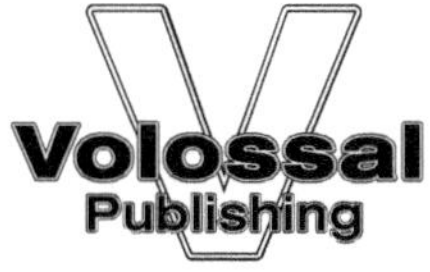

Published by Volossal Publishing
www.volossal.com

ISBN 979-8-9877903-7-3